HOW TO TIME TRAVEL

Explore the Science, Paradoxes, and Evidence

LOUIS A. DEL MONTE

Published by Louis A. Del Monte.
Printed in the United States of America.

ISBN 10: 0988171848
ISBN 13: 9780988171848
Library of Congress Control Number: 2013909805
Louis A Del Monte, Minnetonka, MN

I dedicate this book to the future of humankind. May all that is good in humankind harness the opportunities of time travel and avoid its potential malevolent applications.

Acknowledgments

I would like to acknowledge the support of my wife, Diane Del Monte, without whose help in shouldering the lion's share of the routine aspects related to everyday life, I would not have had the time to write this book. In addition, Diane, who is an artist, provided the central concept for the cover design, as well as many stimulating conversations that helped sharpen my focus in writing the book. I am also grateful to my close friend Nick McGuinness for his chapter-by-chapter suggestions to improve the book. His feedback has been extremely valuable in clarifying and amplifying numerous aspects of this book. Lastly, I am grateful to my colleagues, Anthony Hickl, retired director of technology at a Forbes's privately owned top ten company, and Ed Awad, vice president of the US Division at Dynatek Motion Control, for their endorsements.

Contents

Section III: Building a Time Machine

Introduction

Few subjects evoke more emotion than time travel, the concept of moving between different points in time in a manner analogous to moving between different points in space. Humankind's fascination with time travel dates back thousands of years. Although there is no consensus recognizing which written work was the first to discuss time travel, many scholars argue that the *Mahabharata*, from Hindu mythology, is the first, dating between 700 BCE (Before the Common/Current/Christian Era) and 300 CE (Common/Current/Christian Era). The *Mahabharata*, which is one of the two major Sanskrit epics of ancient India, relates the story of King Revaita, who travels to heaven to meet the deity Brahma. When King Revaita returns to Earth, he is shocked to learn that many ages have passed. In today's science, we would assert King Revaita experienced time dilation.

What is time dilation? It is a scientific fact that time moves slower for any mass accelerated near the speed of light. If that mass were a clock, for example, the hands of the clock would appear to be moving slower than a clock in the hand of an observer at rest. That phenomenon is termed time dilation. If King Revaita used a spaceship capable of speeds near the speed of light to visit Brahma, a roundtrip journey that would appear to King Revaita to take one year would result in a time passage of thirty years on Earth. This may seem like science fiction, but time dilation is a well-established, experimentally verified aspect of Einstein's special theory of relativity; more about this later.

Arguably, the greatest single written work that laid the foundation to fire the imagination of today's generation regarding time travel is H. G. Wells's classic novel, *The Time Machine*, published in 1895. It has inspired numerous popular movies, television programs, novels, and short stories. Why are we humans so obsessed with time travel? It appears to be an innate longing. How many times have you wished that you could go back to a specific point in time and select a different action? We all do it. Consider the number of times you have replayed a specific situation in your mind. Psychologists tell us we replay an event in our minds when the outcome is not finished to our satisfaction. This has accounted for numerous nights of tossing and turning. Another common need is to seek answers to important questions from a firsthand perspective. Perhaps you would like to be a witness during the resurrection of Christ, or be a witness behind the grassy knoll during the Kennedy assassination. Perhaps you miss a loved one who has passed on, and you would like to go back in time to embrace that loved one again.

Some of us also dream about time travel to the future. What outcomes will result from our decisions? Imagine the prosperity and happiness that could be ours if we were able to travel to the future. We would be able to witness the outcome of any decision, return to the present, and guide our lives accordingly. Picking the right profession or choosing the right mate would be a certainty. We could ensure there would be no missteps in our life. A life of leisure and prosperity would be ours for the taking.

It is little wonder that many people ask this deceptively simple question: Is time travel possible? The majority of the scientific community, including myself, says a resounding yes. The theoretical foundation for time travel, based on the solutions to Einstein's equations of relativity, is widely accepted by the scientific community. The next question, which is the most popular question, is how to time travel. Of all the questions in science, the keyword phrase "how to time travel" is close to the top of Internet search engine searches. According to Google, the largest search engine in the world, there are 2,240,000 worldwide monthly searches for the keyword phrase "how to time travel," as of this writing. Unfortunately, it is the most difficult question to answer.

Obviously, interest in time travel is high, and what people want to know most is how to time travel. This high interest, combined with the intriguing real science behind time travel, is what inspired me to write this book.

At this point, I would like to set your expectations. We are going to embark on a marvelous journey. We will examine the real science of time travel, the theoretical foundation that has most of the scientific community united that time travel is possible. We will also examine the obstacles to time travel, and there are many. However, even in the face of all the obstacles, most of the scientific community agrees it is theoretically possible to time travel. The largest issue in time travel is not the theoretical science. It is the engineering. Highly trained theoretical physicists understand the theoretical science of time travel. However, taking the theory and building a time machine capable of human time travel has proved a formidable engineering task. It has not been done, but we are amazingly close. We have already built time machines capable of sending subatomic particles into the future. If you will pardon the pun, it is just a matter of time before we engineer our way through the time travel barrier and enable human time travel.

In setting your expectations, I promise you significant insight into the real science of time travel and an equally incredible insight into the obstacles to time travel. I cannot promise that with this knowledge you will be able to overcome the obstacles and engineer how to time travel. However, you may be the one person destined to harness the science, glean the engineering simplicity, and journey in time. There is only one way to find out, namely, read on.

SECTION I

Time Travel Evidence

"Today, we know that time travel need not be confined to myths, science fiction, Hollywood movies, or even speculation by theoretical physicists. Time travel is possible. For example, an object traveling at high speeds ages more slowly than a stationary object. This means that if you were to travel into outer space and return, moving close to light speed, you could travel thousands of years into the Earth's future."

—**Clifford Pickover**
Time: A Traveler's Guide

"Extraordinary claims require extraordinary evidence."

—**Carl Sagan (1934 – 1996)**

CHAPTER 1

Scientific Evidence that Time Travel Is Real

In this chapter, we are going to discuss the scientific evidence, typically experimental results and highly verified theories, that proves time travel is a scientific reality. We are going to start our discussion with an unusual phenomenon that is likely to shake your grasp on reality. It is twisting the arrow of time.

Twisting the arrow of time

The flow of time, sometimes referred to as the "arrow of time," is a source of debate, especially among physicists. Most physicists argue that time can only move in one direction based on "causality" (i.e., the relationship between cause and effect). The causality argument goes something like this: every event in the future is the result of some cause, another event, in the past. This appears to make perfect sense, and it squares with our everyday experience. However, experiments within the last several years appear to argue reverse causality is possible. Reverse causality means the future can and does influence the past. For example, in reverse causality, the outcome of an experiment is determined by something that occurs after the experiment is done. The future is somehow able to reach into the past and affect it. Are you skeptical? Skepticism is healthy, especially in science. Let us discuss this reverse causality experiment.

In 2009, physicist John Howell of the University of Rochester and his colleagues devised an experiment that involved passing a laser beam through a prism. The experiment also involved a mirror that moved in extremely small increments via its attachment to a motor. When the laser beam was turned on, part of the beam passed through the prism, and part of the beam bounced off the mirror. After the beam was reflected by the mirror, the Howell team used "weak measurements" (i.e., measurement where the measured system is weakly affected by the measurement device) to measure the angle of deflection. With these measurements, the team was able to determine how much the mirror had moved. This part of the experiment is normal, and in no way suggests reverse causality. However, the Howell team took it to the next level, and this changed history, literally. Here is what they did. They set up two gates to make the reflected mirror measurements. After passing the beam through the first gate, the experimenters always made a measurement. After passing it through the second gate, the experimenters measured the beam only a portion of the time. If they chose not to make the measurement at the second gate, the amplitude of the deflected angle initially measured at the first gate was extremely small. If they chose to make the measurement at the second gate, the deflected angle initially measured at the first gate was amplified by a factor of 100. Somehow, the future measurement influenced the amplitude of the initial measurement. Your first instinct may be to consider this an experimental fluke, but it is not. Physicists Onur Hosten and Paul Kwiat, University of Illinois at Urbana-Champaign, using a beam of polarized light, repeated the experiment. Their results indicated an even larger amplification factor, in the order of 10,000.

The above experimental results raise questions about the "arrow of time." It appears that under certain circumstances, the arrow of time can point in either direction, and time can flow in either direction, forward or backward. This is a scientific result, and I am not going to speculate about religious connotations, free will, and the like. Obviously, there are numerous religious connotations possible and a plethora of associated questions.

You may think this experiment is unique in the history of science, but it is not. In chapter 6, we will discuss the classic double-slit experiment. It has baffled scientists for over a hundred years. It also demonstrates that reverse causality is possible.

Theoretical foundations for time travel

Einstein's special and general theories of relativity underpin the science of time travel. They are briefly presented here as theoretical evidence that time travel is real. We will discuss them in detail in section 2, "The Science of Time Travel." In addition, Del Monte's existence equation conjecture is presented as theoretical evidence that time travel is real.

1. Einstein's special theory of relativity—The scientific community considers the special theory of relativity the "gold standard" of scientific theories. It has withstood over one hundred years of experimental verification. In addition to yielding the most iconic scientific equation of all time, $E = mc^2$, it also

5

gave us our first insight into the scientific nature of time and predicted time dilation, both conceptually and mathematically. Time dilation is the experimentally verifiable difference of elapsed time between two events as measured by observers, when either one or both observers are moving relative to each other at a velocity near the speed of light. It is an experimental fact that the second hand on a clock moving at a velocity close to the speed of light moves slower than a clock at rest. Time dilation is real and implies forward time travel. For example, if you board a spacecraft capable of traveling at 650 million miles per hour, a one-day journey measured by a clock onboard the spacecraft would be equivalent to the passage of one year on Earth. Time dilation experiments are routinely performed using particle accelerators, which we will discuss later in this chapter.

2. Einstein's general theory of relativity—Numerous aspects of the general theory of relativity have been verified. For our purposes regarding time travel, it is important to focus on only two:

- Gravitational time dilation—Gravitational time dilation suggests that two observers differently situated from gravitational masses will observe time differently. For example, a clock closer to the Earth will run slower than a clock farther from the Earth. The stronger the gravitational field, the greater the time dilation. This has been experimentally verified using atomic clocks, and we will discuss the results later in this chapter.

- Closed timelike curves—There are numerous solutions to Einstein's equations of general relativity that delineate the world line of a particle is closed, returning to its starting point. In the general theory of relativity, the world line is the path the particle traverses in four-dimensional spacetime. For example, when the particle starts out, it has four coordinates, three dimensional coordinates and one temporal coordinate. Here is a simple analogy. You are in a specific place, definable by three spatial coordinates, reading this book at a specific time, a temporal coordinate. If the world line of a particle returns to its starting point, the particle is said to have returned to its past, suggesting backward time travel is theoretically possible. However, to date, we have not been able to experimentally verify that this aspect of Einstein's general theory of relativity is true. As previously discussed, there is evidence that the "arrow of time" can be twisted, and that events in the future can influence past events. However, this is not conclusive experimental proof that backward time travel is possible.

3. Del Monte's existence equation conjecture—In summary, the existence equation conjecture is derived from Einstein's special theory of relativity and predicts that a mass requires energy to move in time. If additional positive energy is added to the mass, for example, by accelerating it in a particle accelerator and increasing its kinetic energy, the mass will move more slowly in time. I interpret this as the fundamental explanation of time dilation. An interesting aspect of the existence equation conjecture is that it suggests adding negative energy to a mass will cause the mass to move backward in time. Since today's science has been unable to produce and manipulate negative energy, this last point has not been experimentally verified. In chapter 4, we will discuss the existence equation conjecture in

detail. The derivation and experimental validation of the existence equation conjecture is provided in appendices 2 and 3.

Now that we have briefly summarized the theoretical foundations for time travel, let us discuss the experimental results that prove time travel is real.

Evidence of time travel to the future (time dilation)

The material in this section is taken, for the most part, from my book *Unraveling the Universe's Mysteries*, appendix 3.

Velocity time dilation experimental evidence:

Rossi and Hall (1941) compared the population of cosmic-ray-produced muons at the top of a six-thousand-foot-high mountain to muons observed at sea level. A muon is a subatomic particle with a negative charge and about two hundred times more massive than an electron. Muons occur naturally when cosmic rays (energetic-charged subatomic particles, like protons, originating in outer space) interact with the atmosphere. Muons, at rest, disintegrate in about 2×10^{-6} seconds. The mountain chosen by Rossi and Hall was high. The muons should have mostly disintegrated before they reached the ground. Therefore, extremely few muons should have been detected at ground level, versus the top of the mountain. However, their experimental results indicated the muon sample at the base experienced only a moderate reduction. The muons were decaying approximately ten times slower than if they were at rest. They made use of Einstein's time dilation effect to explain this discrepancy. They attributed the muon's high speed, with its associated high kinetic energy, to be dilating time.

In 1963, Frisch and Smith once again confirmed the Rossi and Hall experiment, proving beyond doubt that extremely high kinetic energy prolongs a particle's life.

With the advent of particle accelerators that are capable of moving particles at near light speed, the confirmation of time dilation has become routine. A particle accelerator is a scientific apparatus for accelerating subatomic particles to high velocities by using electric or electromagnetic fields. The largest particle accelerator is the Large Hadron Collider, completed in 2008. Here is an image of one portion of the "ring" that encases the particle being accelerated:

Large Hadron Collider Ring Section

In 1977, J. Bailey and CERN (European Organization for Nuclear Research) colleagues accelerated muons to within 0.9994% of the speed of light and found their lifetime had been extended by 29.3 times their corresponding rest mass lifetime. (Reference: Bailey, J., et al., *Nature* 268, 301 [1977] on muon lifetimes and time dilation.) This experiment confirmed the "twin paradox," whereby a twin makes a journey into space in a near-speed-of-light spaceship and returns home to find he has aged less than his identical twin who stayed on Earth. This means that clocks sent away at near the speed of light and returned near the speed of light to their initial position demonstrate retardation (record less time) with respect to a resting clock.

Gravitational time dilation experimental evidence:

In 1959, Pound and Rebka measured a slight redshift in the frequency of light emitted close to the Earth's surface (where Earth's gravitational field is higher), versus the frequency of light emitted at a distance farther from the Earth's surface. The results they measured were within 10% of those predicted by the gravitational time dilation of general relativity.

In 1964, Pound and Snider performed a similar experiment, and their measurements were within 1% predicted by general relativity.

In 1980, the team of Vessot, Levine, Mattison, Blomberg, Hoffman, Nystrom, Farrel, Decher, Eby, Baugher, Watts, Teuber, and Wills published "Test of Relativistic Gravitation with a Space-Borne Hydrogen Maser," and increased the accuracy of measurement to about 0.01%. In 2010, Chou, Hume, Rosenband, and Wineland published "Optical Clocks and Relativity." This experiment confirmed gravitational time dilation at a height difference of one meter using optical atomic clocks, which are considered the most accurate types of clocks.

Evidence of time travel to the past

Regardless of any reports to the contrary, such as the numerous "proofs" of backward time travel touted as evidence on the Internet, there is no scientific experimental evidence that proves time travel to the past is achievable.

As mentioned previously, we can show via repeatable scientific experiments that the future can reach back into the past and influence experimental results (reference the "Twisting the arrow of time" section in this chapter and "The double-slit experiment" in chapter 6). However, this is not proof of time travel into the past.

What time travel evidence is missing?

Three crucial pieces of scientific evidence are missing to make an irrefutable case that human time travel is scientifically possible to the past, as well as the future. They are:

1. Although we have numerous experiments that demonstrate time dilation (i.e., forward time travel) involving subatomic particles is real, we have been unable to demonstrate significant human time dilation. By the word "significant," I mean that it would be noticeable to the humans and other observers involved. To date, some humans, such as astronauts and cosmonauts, have experienced forward time travel (i.e., time dilation) in the order of approximately 1/50th of a second, which is not noticeable to our human senses. If it were in the order of seconds or minutes, then it would be noticeable. Therefore, the first missing evidence is significant human forward time travel.

2. There is no experimental evidence that proves we can send even a subatomic particle backward in time. While the

solutions to the equations of general relativity showing closed timelike curves give us a substantial theoretical foundation for backward time travel, we have no experimental evidence that confirms this. Until we can send at least a subatomic particle backward in time, the solutions of general relativity showing closed timelike curves (i.e., backward time travel) will remain a mathematical curiosity. There is no doubt in my mind that any successful backward time travel experiment would be a monumental breakthrough for the science of time travel, and likely qualify for a Nobel Prize.

3. We do not have any scientifically verified evidence that we have been visited by time travelers. In the next chapter, I will present anecdotal evidence that appears to indicate we have been visited by time travelers. However, anecdotal evidence is not scientific evidence. The fact remains, no scientific evidence of time travelers from the future (or any time frame) exists.

Conclusions from Chapter 1

- The "arrow of time" is not rigidly set to point in only one direction, from the present to the future, as most philosophies hold. Scientifically, recent experiments, delineated above, say it is possible to twist the arrow of time and have it point from the future to the present.

- Forward time travel evidence exists. Numerous time dilation experiments involving subatomic particles prove that time dilation (i.e., forward time travel) is real. However, we have been unable to demonstrate significant human time dilation.

- There is no scientific experimental evidence that proves time travel to the past is achievable.

- There is no scientific evidence that we have been visited by time travelers from the future (or any time frame).

This chapter provided scientific evidence that time travel to the future is real. The scientific community can embrace it. It finds its way into formal and prestigious scientific journals. We have no experimental evidence that backward time travel is real. Many in the scientific community greet research on backward time travel with skepticism. This is especially true of human time travel to the past. However, in the next chapter, we will look at some anecdotal evidence that suggest human time travel to the past is not only possible, but is already occurring.

"What would happen if history could be rewritten as casually as erasing a blackboard? Our past would be like the shifting sands at the seashore, constantly blown this way or that by the slightest breeze. History would be constantly changing every time someone spun the dial of a time machine and blundered his or her way into the past. History, as we know it, would be impossible. It would cease to exist."

—**Michio Kaku** *Hyperspace*

CHAPTER 2

Anecdotal Evidence that Time Travel Occurs

You can find information on almost anything you want to know. Before the Internet, the major metropolitan libraries held the keys to research. Today, almost any information is one click away on the Internet. For example, it is even possible to learn how to build a nuclear weapon, just using your computer and a search engine.

However, while gathering information has been simplified, distilling it into useful intelligence is hard. It is much like a jigsaw puzzle. It starts with just one piece of information and connecting it to another piece, until a meaningful picture emerges. Each piece of information, in and of itself, may appear useless. However, if you gather enough information, piece by piece, almost inevitably, a picture emerges. Interpretation remains subjective and open to error, but it is amazing what you can piece together from Internet and literary research.

Let me give you an example. When I was a child, one of my hobbies was building small plastic models, including model planes, ships, and the like. In 1954, when I was ten years old, the United States Navy launched the world's first atomic submarine, the USS *Nautilus* (SSN-571). It made headlines throughout the world. Everybody was talking about it, and every model builder wanted to build a model of it. With that level of demand, Revell Hobby Kits offered a plastic model build kit. Revell's model of the USS *Nautilus* (SSN-571) was so accurate it started people speculating. Did Revell somehow get their hands on the blueprints of the *Nautilus*? Apparently, the Revell model engineers knew how to do research, even without the Internet. Surprisingly, they were able to fill in the blanks and pinpoint the almost exact location of the atomic reactors on the *Nautilus*, which at the time was top secret. It is a childhood memory that impressed me so deeply, it is still with me to this day, and I can remember talking about it with my friends. We thought it was funny at the time. It probably did not amuse the United States Navy.

I have spent countless hours researching this book. This is a book on the real science of time travel, which rests on Einstein's special and general theories of relativity, and on the existence equation conjecture. Governments from Nazi Germany to the United States have been pursuing technologies that would enable time travel. For example, Nazi Germany reportedly pursued antigravity experiments via "Die Glocke" (German for "The Bell"), which was first described by Polish journalist and author Igor Witkowski (2000). Later, military journalist and author Nick Cook, as well as writers Joseph P. Farrell and Jeremy Robinson, wrote about it and popularized it as a Nazi secret weapon antigravity device. The United States Air Force funded work on creating and storing antimatter, as far back as the 1960s (Reference: "Air Force pursuing antimatter weapons/Program was touted publicly, then came official gag order," by Keay Davidson, *San Francisco Chronicle*, Monday, October 4, 2004).

Why would governments be interested in time travel? When you think about it, time travel would be the ultimate weapon. With it, history would be rewritable at will. Simply time travel to the past and change the outcome of an event. Spies, secret agents, spy satellites, and the like would become obsolete. Simply time travel to the future and see what new capabilities and weapons an enemy has. Then time travel back to the present and report.

Governments are interested in the technologies associated with time travel. Public records indicate they have been pursuing time-travel-related technologies for a half century or more. However, how successful has this pursuit been? Have we been able to time travel?

To address the above questions, let us examine the available evidence from the Internet, the Bible, and other sources. However, here is a word of caution. **This evidence is not scientific fact.** It is anecdotal evidence, and it may even be bogus. Nonetheless, the sheer volume of anecdotal evidence makes it hard to ignore. Will we be able to take each piece of evidence and distill it into a compelling scenario, a mosaic of the truth? You be the judge.

Evidence from the Internet

If you do an Internet search with Google using the keyword phase "time travel evidence" (without the quotes), you will get about

297,000,000 search returns. Most of the evidence falls into three categories:

1. Old movie clips: There are a number of YouTube videos of old movies showing people using devices, such as a cell phone, that would not have existed when the movie was made. One example that comes up numerous times is Charlie Chaplin's 1928 film *The Circus,* featuring a woman who appears to be talking on her cell phone. Just do a YouTube search using the keyword phrase "time travel evidence Chaplin film" (without the quotes). You will get back over a 100,000 search returns. The first page or two of search results have clips of this video, typically with some commentary. Here is a still picture taken from the video clip:

Video clip of a woman supposedly using a cell phone in the 1928 Charlie Chaplin film *The Circus*

2. Old photographs: Many sites include old photographs that show people out of context, for example, wearing clothing that does not fit the time, such as modern sunglasses, or using devices, such as a 35mm camera, that did not exist at the time the photograph was taken. To see these results do a Google search using the phrase "time travel photo

evidence" (without the quotes). One website that has a number of good examples of this is http://www.trutv.com/ conspiracy/paranormal/time-travelers/gallery.all.html. It features a 1941 photograph of a person with Ray-Ban sunglasses and a 35mm camera. Here is the photo of the man:

Man with Ray-Ban sunglasses, holding a 35mm camera

3. Archeological finds: There are archaeological finds of modern devices, such as a one-hundred-year-old Swiss-made watch found in a four-hundred-year-old Ming dynasty tomb in Shangsi County, Guangxi, in southern China. If you do a Google search using the keyword phrase "time travel watch evidence" (without the quotes) you will get about 74,000,000 search results. Here is a photograph of the ring watch that was supposedly found in the four-hundred-year-old Chinese tomb:

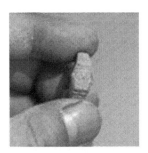

Ring Watch

There is no lack of anecdotal evidence of time travel on the Internet. However, we need to ask some serious questions. Is the evidence real?

Are we reading more into the evidence than is warranted? Is it a hoax? I think these are all legitimate questions. Here are some answers, but, much like the evidence, they are not scientifically verifiable.

Let us start with the 1928 Charlie Chaplin film, *The Circus*. Debunkers argue that the woman was just holding a primitive hearing aid known as an ear trumpet. Here is a photograph of a 1928-era ear trumpet:

1928-era Ear Trumpet

Surprisingly, it looks like a cell phone from a distance. Proponents dismiss this as an explanation because the woman is talking into it. However, you see some people talking aloud to themselves all the time. This does not mean they are crazy. This is just how they process information and think. Almost all of us talk to ourselves privately in our minds. It is called thinking. Does this explain the film clip?

Let us next examine some of the photographic evidence. I have made two observations:

1. Many of the old photographs are fuzzy. This is typical of old photographs, since cameras in the early part of the twentieth century were crude.

2. The claims that something or someone is "out of context" are a bit of a stretch. For example, consider the man in the 1941 photograph. Some suggest he is wearing Ray-Ban sunglasses and a screen-print T-shirt, and holding a modern 35mm

camera. I think the photograph is too fuzzy to make a solid case for these assertions, but that is just my opinion. I suggest you view the photograph and draw your own conclusion.

In addition, with today's computer technology and state-of-the-art photograph-editing programs, such as Photoshop, it is possible to manipulate a photograph and have Elvis shaking hands with Albert Einstein. Only a highly trained computer photographic expert would be able to determine that the photograph is a computer-generated manipulation of pixels—in other words, a fake. The technology is that good. This makes me suspicious of all photographic evidence that has not been analyzed by a highly trained expert.

Lastly, let us examine the archaeological finds, specifically the one-hundred-year-old Swiss-made watch found in the four-hundred-year-old Ming dynasty tomb in Shangsi County, Guangxi, in southern China. Several aspects of this suggest it is a hoax. First, if the watch is supposed to be one hundred years old, it does not fit with historical facts. According to historians, the watch does not resemble watches of the era (i.e., this watch supposedly dates back to about the early twentieth century). Secondly, in the writing on the back of the watch, the word "Swiss" is in English. This does not fit the period of an early twentieth century Swiss-made watch. It would be more likely for the word to be in German or French, similar to the writing on other watches of the period. Does this prove it is a hoax? No! However, it provides a reason to doubt its authenticity.

Where does this leave us regarding time travel evidence from the Internet? In a phrase, it leaves us on "shaky ground." I examined only some of the most popular time travel evidence on the Internet, and it is far from conclusive. Evidence that is more conclusive may lie buried in the 297,000,000 Google search returns for the keyword phrase "time travel evidence." The challenge is finding it. A true scholarly effort would likely take a lifetime. However, in my opinion, the real world, and the universe, is stranger than any work of fiction. Therefore, I am keeping an open mind, and I suggest you do the same.

Evidence from urban legends

A time travel urban legend is usually a story about someone or something that has supposedly traveled in time. Before the Internet, the urban legends went viral mouth to mouth. After the Internet, they went viral mouse to mouse. In addition, many urban legends owe their popularity to the press, to popular books on the subject, or to becoming the subject of one or more motion pictures.

The list of time travel urban legends is far too large to cover them all. I have chosen what I consider the most popular urban legend, namely, the Philadelphia Experiment, to discuss in detail. In my judgment, it is a good representation of the urban legend category. Word of mouth, books, the Internet, motion pictures, and documentaries have popularized it, and it is still making the rounds alive and well today.

The Philadelphia Experiment, also known as Project Rainbow, allegedly had the objective of "cloaking" (i.e., rendering invisible) the United States Navy destroyer escort *Eldridge*, shown in a 1943-era photograph.

1943-era Photograph of USS Eldridge

Purportedly, in the process of cloaking the *Eldridge*, strange phenomena occurred, including time travel. The experiment supposedly took place at the Philadelphia Naval Shipyard, on or around October 28, 1943.

Like many urban legends, there are numerous accounts of the Philadelphia Experiment. What follows are key story points common to most accounts. Allegedly, the experiment had full navy backing and was based on unified field theory (a term coined by Albert Einstein), which seeks to unite the fields of electromagnetism (in this case, light) and gravity into a single field. The theory was to bend light around the ship using large electrical generators, consequently bending spacetime and rendering the ship an invisible time machine.

There are no reliable attributable accounts, but supposedly, the *Eldridge* was fitted with the electrical generators in the summer of 1943 by "researchers," whose identity remains unknown. After being properly equipped, testing began, reportedly with some success. Here are the salient test accounts:

- July 22, 1943—The *Eldridge* was rendered invisible, some witnesses reporting a "greenish fog" in its place.

- October 28, 1943—The *Eldridge* vanished in a flash of blue light and teleported to Norfolk, Virginia, about two hundred miles away. The *Eldridge* sat in full view of men aboard the SS *Andrew Furuseth*, a nearby merchant ship, for an unspecified period of time, whereupon the *Eldridge* vanished and reappeared at the original Philadelphia site, traveling approximately ten seconds (in some accounts longer) back in time.

According to many accounts, the experiments caused the crew to experience serious side effects. A number of accounts claim some members of the crew were fused physically to the metal structures of the ship, the atoms of their bodies intermixed with the atoms of the ship. For example, one sailor supposedly had his hand embedded in the steel hull of the ship one level below where he was originally standing. Others crew members were said to suffer nausea and mental disorders, and even to vanish completely. To make the story complete, the navy is said to have "brainwashed" any *Eldridge* survivors to prevent them from revealing the incidents.

The Philadelphia Experiment is as good as urban legends get, supposedly incorporating the science of Einstein, a government secret experiment, unexplainable phenomena, and brainwashed survivors. It is only natural to ask: How did the Philadelphia Experiment urban legend get started?

The origin of the Philadelphia Experiment urban legend is itself another urban legend. We have one urban legend underpinning another. To quote Winston Churchill, we have *"a riddle, wrapped in a mystery, inside an enigma."* Again, there are numerous accounts of the origin of the Philadelphia Experiment. We will use the 2002 book by James Moseley and Karl Pflock, *Shockingly Close to the Truth!: Confessions of a Grave-Robbing Ufologist*, as a reference on the origin of the Philadelphia Experiment urban legend.

According to Moseley and Pflock's 2002 book, in 1957, the Office of Naval Research (ONR) in Washington, DC, contacted Morris K. Jessup, an astronomer and author of the 1955 book *The Case for the UFO*. He was asked to study the contents of a parcel that the ONR had received in a manila envelope marked "Happy Easter." It was a paperback copy of Jessup's UFO book that had been extensively annotated in its margins. Moseley and Pflock claim that annotations were written with three different shades of pink ink. The annotations detail correspondences among three individuals: "Jemi," and two others the ONR labeled "Mr. A." and "Mr. B." The annotators refer to themselves as "Gypsies," discuss people living in outer space, and comment on the merits of Jessup's assumptions in the book. The annotations also contain a reference to the Philadelphia Experiment.

The ONR asked Jessup if he knew anything about the annotations, including knowledge of those involved. Jessup identified "Mr. A" as Carlos Allende. Supposedly, Allende had sent Jessup a letter in 1955 claiming to have served on the SS *Andrew Furuseth* and claiming to have direct knowledge of the Philadelphia Experiment. Allende claimed he witnessed the *Eldridge* appear and disappear. When Jessup requested that Allende provide evidence and corroboration, Jessup received another correspondence. This new correspondence came from a man identifying himself as Carl M. Allen. Allen said that he could not provide the evidence and corroboration Jessup sought. However, Allen implied that he might be able to recall some details via hypnosis. This all seemed highly suspicious to Jessup, and he discontinued the correspondence. (Apparently, Jessup's suspicions were well founded. The ONR determined the return address on Allende's letter to Jessup was an abandoned farmhouse.)

Now, the story becomes even stranger. According to Moseley and Pflock's 2002 book, the ONR decided to fund a small printing, about a hundred copies, of the annotated volume, complete with both letters Jessup had received from Allende/Allen. The Texas-based Varo Manufacturing Company did the printing. Supposedly, the ONR gave Jessup three copies and circulated the rest within the navy. For those interested, I found a copy of the Varo edition online at this website: http://obscurantist.com/files/case-for-ufos-annotated.pdf.

Jessup began to write extensively on the topic in an attempt to make a living, but his follow-up book did not sell well, and the publisher rejected his other manuscripts. Jessup became depressed, and his life took a turn for the worse when he was involved in a car accident. This further added to his depression, and Jessup committed suicide on April 20, 1959.

What gives this urban legend legs are three published books. These are not the only books on the Philadelphia Experiment. However, according to historian Mike Dash, numerous authors appear to take their information from one of the three sources below:

1. Jessup's 1955 book, *The Case for the UFO*

2. Moseley and Pflock's 2002 book, *Shockingly Close to the Truth!: Confessions of a Grave-Robbing Ufologist*

3. The ONR's Varo edition of Jessup's book, complete with annotations and letters Jessup had received from Allende/Allen

However, is any of it true? Do the facts support any portion of the Philadelphia Experiment?

In 1980, Robert Goerman wrote in *Fate* magazine that Carlos Allende/Carl Allen was Carl Meredith Allen of New Kensington, Pennsylvania. Carl Meredith Allen had a history of psychiatric illness. Goerman speculates that Allen may have fabricated the Philadelphia Experiment as a result of his mental illness. Later Goerman characterized Allen as "a creative and imaginative loner...sending bizarre writings and claims."

Berlitz and Moore's book *The Philadelphia Experiment: Project Invisibility*, published in 1979, claims to include factual information, such as an interview with a scientist involved in the experiment. This book is considered a definitive source for information on the Philadelphia Experiment. However, some critics accuse Berlitz and Moore of plagiarizing story elements from the novel *Thin Air*, by George E. Simpson and Neal R. Burger, which was published a year earlier. However, this criticism may not be fair. Earlier works on the subject likely inspired *Thin Air*, including Berlitz's chapter on the experiment in his 1977 book, *Without a Trace: New Information from the Triangle*. This suggests the criticism of plagiarism is unwarranted. Berlitz and Moore's book may be the real deal.

Now, let us discuss the science and other facts surrounding the Philadelphia Experiment. From the standpoint of science, light bends, in accordance with Einstein's general theory of relativity, when it is near the surface of an extremely massive object, such as a sun or a black hole. No known or published scientific apparatus exists that enables us to bend light around an object the size of a navy ship. Could the navy have secretly developed such an apparatus by 1943? I do not think it is likely, but I will not rule it out altogether. The science claimed to be used in the Philadelphia Experiment was unified field theory. Factually, even today, there is no accepted unified field theory, but it is an area of ongoing research. Einstein was working on unified field theory, attempting to unify electromagnetism with general relativity, his theory of gravity. Some accounts of the Philadelphia Experiment suggest that Einstein was successful, but chose not to publish it.

The USS *Eldridge* was commissioned on August 27, 1943. This is one month after the first experiment was reported to occur. According to official records, it remained in port in New York City until September 1943. Also according to official records, the *Eldridge* was on its first shakedown cruise in the Bahamas during the time the October experiment was reported to occur. Proponents of the Philadelphia Experiment argue that the ship's logs have been falsified and the real logs are classified.

In 1996, the Office of Naval Research (ONR) stated, "ONR has never conducted investigations on radar invisibility, either in 1943 or at any other time." In addition, they pointed out that the ONR was not established until 1946, three years after the Philadelphia Experiment. The implication is the ONR did not exist or conduct the Philadelphia Experiment. Further, the ONR denounced the Philadelphia Experiment as "science fiction." This appears to be corroborated by the navy veterans who served aboard the USS *Eldridge*. During a 1999 reunion, the *Eldridge* veterans told a Philadelphia newspaper that their ship had never made port in Philadelphia.

Some critics debunk the Philadelphia Experiment by arguing it was just a case of misinterpretation. They argue the *Eldridge* was "degaussed" (the process of making a steel ship's hull nonmagnetic) while in port, and this procedure started the urban legend. It is a fact that the *Eldridge* was degaussed. This was a common procedure to render a ship undetectable to magnetically fused undersea mines and torpedoes. It required the generation of a strong electromagnetic field onboard the ship. Charles F. Goodeve invented this procedure when he was a commander in the Royal Canadian Naval Volunteer Reserve. The Royal Navy and United States Navy used it widely during World War II.

Is the Philadelphia Experiment fact or fiction? Is it possible the United States Navy has been able to orchestrate a consistent set of lies over a period of what is now about seventy years? You will have to be the judge. I have provided the story points proponents and opponents of the Philadelphia Experiment cite. Which side do you favor? Regardless of which way you lean, one thing is certain. The accounts of the Philadelphia Experiment are intriguing, and they are still making the rounds over seventy years since the alleged first experiment. It is, in my opinion, representative of the category of urban legends related to time travel.

Evidence from the Bible

To be clear, I am not citing the Bible in this section as a sacred text of Judaism or Christianity. The Bible is a collection of writings, having numerous versions, which some scholars believe have roots that date back thirty-five hundred years. Numerous ancient texts tell stories of time travel. I am citing the Bible as a reference, representative of ancient texts. I chose the Bible for two reasons:

1. In its numerous versions, the Bible is the most widely published book of all time.

2. The Bible carries high credibility with a significant segment of Americans. According to the Gallup News Service, a poll published in 2007 stated, "About one-third of the American adult population believes the Bible is the actual word of God and is to be taken literally word for word."

I intend to present only one time travel story from the Bible. I judge it representative of biblical time travel stories. It has stirred controversy for centuries. It is commonly known as "Joshua's long day." The context of the story relates to when the Israelite army was involved in the conquest of Canaan. The biblical scripture suggests that God supernaturally intervened by lengthening the daylight period of a particular battle day, thus allowing the Hebrews to achieve a great victory. There are many versions of the story, depending on the Bible edition chosen. This version is from the King James 2000 Bible (copyright 2003), Joshua 10:12–14:

> 12 *Then spoke Joshua to the LORD in the day when the LORD delivered up the Amorites before the children of Israel, and he said in the sight of Israel, Sun, stand you still upon Gibeon; and you, Moon, in the valley of Aijalon.*
>
> 13 *And the sun stood still, and the moon stayed, until the people had avenged themselves upon their enemies. Is not this written in the book of Jasher? So the sun stood still in the midst of heaven, and hastened not to go down about a whole day.*
>
> 14 *And there was no day like that before it or after it, that the LORD hearkened unto the voice of a man: for the LORD fought for Israel.*

What does this mean? If we take it literally, time significantly slowed down to the point that about a day was lost. In a sense, this is a time dilation story.

However, did it literally happen? Broadly, there appear to be two camps on the issue:

1. The science camp, which claims that Joshua's long day is scientifically provable.

2. The fiction camp, which claims the story is fiction, intended not to be taken literally, but to be understood within a religious context.

The science camp appears to base their arguments on Baltimore industrialist Harold Hill's October 1969 article in *Evening World*, a newspaper in Spencer, Indiana. The 1969 article was picked up by numerous Christian publications. In 1974, Hill again published the account in his book *How to Live Like a King's Kid*, which is still being sold today. According to the Hill account, NASA "space scientists were checking the position of the sun, moon, and planets out in space, calculating where they would be 100 and 1,000 years from now," when the computers came to a grinding halt. Apparently, according to Hill, the computers found an anomaly related to a missing day. As the story goes, one of the space scientists suggested this could be related to Joshua's long day.

This is an intriguing story, but is it factual? According to an article published August 29, 2005, by Dr. Bryant G. Wood at biblearchaeology. org, "Joshua's 'Long Day' and Mesopotamian Celestial Omen Texts," it is false. Wood claims he contacted NASA in May 1979 and got this response from Edward Mason, chief, Office of Public Affairs: *"We know nothing of Mr. Harold Hill and in no way can corroborate the 'lost day' reference in the article."* At this point, Wood claims he wrote to Hill in May 1970. Hill replied with a form letter that said:

> *Since this incident took place about two years ago I have misplaced the source information and so am unable to give you*

names and places but will send it to you when I locate it. In the meantime I can only tell you that had I not considered the source to be completely reliable I would not have made use of it in the first place.

No source information was ever provided by Harold Hill, who died in 1987. Hill was president of the Curtis Engine and Equipment Company in Baltimore for twenty-two years. The company had this disclaimer on their website in or around 2005, but it has since been removed:

Curtis Engine & Equipment, Inc., does not substantiate or continue Harold's personal research. Unfortunately, we have no information concerning the "Missing Day" nor do we have any connections at NASA who would be able to corroborate Harold's claims.

There are a number of accounts on the Internet offering scientific explanations for Joshua's long day. In fact, far too many to list or discuss in these pages. For the most part, I judge that the NASA response says it all. In my opinion, NASA would have issued a press release had they found a missing day.

Does this mean Joshua's long day did not occur? No! There are similar accounts of a long day or a long night in other texts throughout the world. Internet searches will reveal tales of either a long night or a long day, dependent on world location, in North America, Central and South Americas, China, and Africa, just to name a few. Are they concurrent with Joshua's long day? That is hard to prove. Nailing down the exact date of Joshua's long day is difficult. Correlating it with the other tales of long days and nights would be a formidable task, if even possible. Many elements of the Bible are inspired by actual real-world events. I think we need to accept the possibility that it could have happened. The scientific reasons explaining it have not garnered a supportive audience in the scientific community. However, this does not mean the long day is false. Einstein's special theory of relativity, published in 1905, did not garner a supportive audience in the scientific community when it was first published. In approximately twenty years, it was accepted worldwide. What this says to me is that we need to maintain an open mind, but require that scientific data be verifiable.

Let us turn our attention to the Joshua's long day "fiction camp." The fiction camp does not take Joshua's long day literally. They argue that the biblical author is attempting to make a religious point, not a scientific point. This makes the most sense to me. Biblical scholars argue the Bible is not to be taken literally, but rather interpreted to understand the religious message the biblical author intended. This unshackles it from any scientific interpretations, which are numerous and highly controversial. I am intentionally not going to comment on its religious significance, since this is a book on science.

There you have it. One time travel story, actually a time dilation story, from the Bible. Is it real? Did Joshua's long day actually occur? You have the salient facts. I think that it is up to you to judge.

Joshua's long day is one of many time travel accounts in the Bible. If biblical time travel stories are of high interest to you, there is a book dedicated to the subject by Gary Stearman, *Time Travelers of the Bible: How Hebrew Prophets Shattered the Barriers of Time-Space* (published 2011). It is available at Amazon.com.

Evidence from UFOs

Internet searches for the keyword acronym "UFO" (unidentified flying object) are among the most popular on the Internet. According to Google, there are five million global searches per month for the keyword acronym "UFO" (without the quotes).

Let us start with a little background. Surprisingly, the United States Air Force (USAF) officially created the acronym "UFO" in 1953. Their intent was to replace the more popular phrases such as "flying saucers" and "flying discs" because of the variety of shapes reported. In their official statement, the United States Air Force defined the term UFO as "any airborne object which, by performance, aerodynamic characteristics, or unusual features, does not conform to any presently known aircraft or missile type, or which cannot be positively identified as a familiar object."

The phenomena, namely UFO sightings, are worldwide. Various governments and civilian committees have studied them. The conclusions reached by the various organizations that have studied them vary significantly. Some conclude UFOs do not represent a threat and are of no scientific value (see, e.g., 1953 CIA Robertson Panel, USAF Project Blue Book, Condon Committee). Others conclude the exact opposite (see, e.g., 1999 French COMETA study, 1948 USAF Estimate of the Situation, Sturrock Panel).

Given the sheer volume of unexplained sightings by credible witnesses, including military, police, and civilian witnesses, there is little doubt that the UFO phenomenon is real and worldwide, and for the most part, there is no widely accepted public or scientific explanation of what they are or what their intentions might be.

Three popular speculations regarding UFOs are:

1. They are future generations of humans who have mastered the science of time travel, and they are coming back either to observe us or to carry out other intentions.

2. They are technologically advanced aliens from another planet who have mastered the science of time travel, and they are coming here either to observe us or to carry out other intentions.

3. They are secret government (United States or any government) experimental spacecraft, and by some accounts they are reverse engineered from advanced alien spacecraft in the government's possession.

In my estimation, the ninety-page 1999 French COMETA study (the English translation stands for "Committee for In-Depth Studies") is the most authoritative source of UFO information and provides a thoughtful, balanced view. Here are the facts that led me to this position:

- The COMETA membership consisted of an independent group of mostly former "auditors" (i.e., defense and intelligence analysts) at the Institute of Advanced Studies for National Defense, or IHEDN, a high-level French military think tank, and by various other highly qualified experts. The independence of the group lends credence that the findings and conclusions would not be censored.

- The French government did not sponsor it. This lends credence that the COMETA members were objective and not politically guided.

- The COMETA study was carried out over several years. This lends credence that the COMETA study is a thorough account of UFO phenomena, not a hastily put out government press release.

The 1999 COMETA study concluded:

1. About 5% of the UFO cases studied were inexplicable.

2. The best hypothesis to explain them was the extraterrestrial hypothesis (ETH), but they acknowledged this is not the only possible hypothesis.

3. The authors accused the US government of engaging in a massive cover-up of UFO evidence.

According to the 1999 COMETA study, a small but significant percentage of UFOs are likely of extraterrestrial origin. Does this rule out that they are future generations of humans, visiting the past? In my opinion, it does not. Even the conclusions of the 1999 COMETA study did not rule out this possibility. However, there is no conclusive evidence either way.

You will find an English translation of the 1999 COMETA study at this website address: http://www.ufoevidence.org/newsite/files/COMETA_part2.pdf.

The main questions regarding time travel and UFOs are:

- Are the UFOs future generations of humans, time traveling back to our past and present?

- Are the UFOs alien spacecraft, or secret government experimental spacecraft, able to traverse great distances using technologies essential to time travel, like a matter-antimatter propulsion system?

I suggest you read the complete 1999 COMETA study and draw your own conclusions.

Assessing the scientific validity of the anecdotal evidence

By its nature, anecdotal evidence of time travel is not scientifically valid. However, based on the sheer volume of anecdotal evidence (validated by simple Google searches), I judge some percentage of it may be valid. Even if that percentage is small, the implications are immense. Here are some salient points to consider:

- Because of the science and energy required, any time travel is likely under the control of governments, not individuals. This implies that time travelers are government agents.

- If time travelers visit us, they may be under strict directives regarding nonintervention, for numerous reasons, including the unknown effects of time travel paradoxes.

- If UFOs are of extraterrestrial origin, they obviously have mastered either faster-than-light space travel or have learned to develop and use traversable wormholes. In other words, they have mastered the science of time travel.

- If UFOs are secret government experimental spacecraft, this suggests a government (the United States or any government) has mastered some of the technologies that would be essential to time travel.

- Keeping secrets is difficult. Just browse the website http://wikileaks.org. Keeping military secrets over a prolonged period is extremely difficult. In the words of Benjamin Franklin, in *Poor Richard's Almanack*, "Three may keep a secret, if two of them are dead." Are the time travel incidences we read on the Internet and elsewhere the result of leaked secrets?

Conclusions from Chapter 2

- There is a huge volume of anecdotal evidence of time travel on the Internet, in the Bible, and in the public record (i.e., newspapers, books, etc.).

- The sheer volume implies that some may be valid, but by the nature of anecdotal evidence, this is impossible to prove scientifically.

- The 1999 French COMETA study is the most authoritative source of UFO information. It makes a compelling case that about 5% of UFO sightings are genuine. If true, this lends credence to three possible scenarios:

    1. They are future generations of humans who have mastered the science of time travel, and they are coming back either to observe us or to carry out other intentions.

    2. They are technologically advanced aliens from another planet who have mastered the science of time travel, and they are coming here either to observe us or to carry out other intentions.

    3. They are secret government experimental spacecraft that demonstrate one or more governments have mastered some of the technologies essential to time travel.

The Science of Time Travel

"People like us, who believe in physics, know that the distinction between past, present, and future is only a stubbornly persistent illusion."

—**Albert Einstein (1879 – 1955)**

"Time travel was once considered scientific heresy. I used to avoid talking about it for fear of being labelled a crank. But these days I'm not so cautious. In fact, I'm more like the people who built Stonehenge. I'm obsessed by time."

—Stephen Hawking (1942 –)

CHAPTER 3

What Is the Science of Time Travel?

The science of time travel is real. There is experimental evidence that proves time travel is real. Yet, with but a few exceptions, most of my colleagues in the scientific community avoid discussing or doing serious time travel research. Why is this?

The theory regarding time travel is relatively easy to understand on a technical basis if you have or are pursuing a degree in the physical sciences, or on a conceptual basis, for the layperson. For example, professors teach time dilation (i.e., forward time travel) in undergraduate physics classes. Professors also teach general relativity in both undergraduate and graduate physics classes. The general theory of relativity embodies, along with Einstein's theory of gravity, the science of time travel to the past. Both the special and general theories of relativity are easy to grasp for a person with the proper scientific background. However, designing and engineering experiments to demonstrate time travel is an extremely difficult task. In fact, building particle accelerators capable of demonstrating even the simplest form of time travel, time dilation, requires the participation of numerous institutions, numerous nations, and a huge financial investment. An example of this is the Large Hadron Collider (LHC), which is the world's largest high-energy particle accelerator. The European Organization for Nuclear Research (CERN), a collaboration of ten thousand scientists and engineers from over one hundred countries, built the LHC over a ten-year period, 1998 to 2008, at an estimated cost of $9 billion. Scientists hail it as one of the greatest scientific achievements. It is able to perform time dilation experiments, among many other important scientific tasks. However, even with highly sophisticated scientific instruments, research regarding particle acceleration and detection is a difficult endeavor. For example, in 2011, scientists using the Oscillation Project with Emulsion-tRacking Apparatus (OPERA) reported accelerating neutrinos faster than the speed of light, which later proved incorrect and due to faulty cable connections. In 2012, CERN scientists announced they had confirmed the existence of the Higgs boson, which made headlines throughout the world. However, contrary to the news reports, the confirmation of the Higgs boson is still in question and is not universally accepted. The main point is that the apparatus proposed to perform time travel research, even using subatomic particles, is extraordinarily expensive, difficult to build, and difficult to use. The energy required, even when dealing with subatomic particles, is enormous. We will revisit the question regarding the amount and type of energy required for time travel in the next chapter.

This chapter is devoted to the theoretical science of time travel. We will start our understanding of this subject by exploring Einstein's special theory of relativity.

Einstein's special theory of relativity (forward time travel)

Einstein's position as a patent examiner at the Federal Office for Intellectual Property, in Bern, Switzerland, must have afforded him a lot of time to think. While working as patent examiner in 1905, the twenty-six-year-old Albert Einstein received his PhD from the University of Zurich and published four scientific papers in the prestigious *Annalen der Physik* (i.e., *Annals of Physics*), one of the oldest scientific journals, which was first published in 1790. The papers, originally handwritten, discussed the photoelectric effect, Brownian motion, special relativity, and the equivalence of mass and energy. Unfortunately, few of Einstein's contemporaries understood the conceptual significance of his publications. However, that was about to change over the next fifteen years, and the world of physics would never be the same again.

In this section, we are only going to explore Einstein's special theory of relativity, for it is this theory that lays the foundation of time travel to the future. The paper that Einstein submitted regarding his special theory of relativity was titled "On the Electrodynamics of Moving Bodies." By scientific standards, it was unconventional. It contained little in the way of mathematical formulations or scientific references. Instead, it was written in a conversational style using thought experiments. If you examine the historical context, Einstein had few colleagues in the scientific establishment to bounce ideas off. In fact, Einstein essentially cofounded, along with mathematician Conrad Habicht and close friend Maurice Solovine, a small discussion group, the Olympia Academy, which met on a routine basis at Solovine's flat to discuss science and philosophy. It is also interesting to note that Einstein's position as a patent examiner related to questions about transmission of electric signals and electrical-mechanical synchronization of time. Most historians credit Einstein's early work as a patent examiner with laying the foundation for his thought experiments on the nature of light and the integration of space and time (i.e., spacetime).

Before we undertake understanding Einstein's special theory of relativity, let me set your mind at rest. This will be primarily a conceptual discussion. The objective is to provide the reader with an appreciation of special relativity, along with its implications to the science of time travel.

Einstein formulated his special theory of relativity based on two postulates:

1. The laws of physics remain constant in all inertial frames of reference. An inertial frame of reference is any frame of reference moving at a constant velocity. For example, if you are in your car traveling down the highway at a constant speed of sixty miles per hour, you are in an inertial frame of reference. Einstein termed this the principle of relativity. The implication is that any physical experiment done in one inertial frame of reference is indistinguishable from the same experiment done in another inertial frame of reference. Therefore, if you strike a match to light a cigarette in a car going sixty miles per hour, the effect (striking the match and lighting the cigarette) would be the same in a car going thirty miles per hour, ignoring outside influences like an open window. Historically, Einstein's principle of relativity has its roots in Galileo's principle of relativity, in which he postulated that all motion was relative and there was no absolute frame of rest. Einstein adopted Galileo's principle of relativity, postulated that it held for all the laws of physics, and extended it to include light, as will become evident in the second postulate of relativity.

2. The speed of light is a constant in all inertial frames of reference. This may seem counterintuitive. If you were in a car moving at a constant velocity, for example, sixty miles per hour, and threw a ball thirty miles per hour in the same direction the car was moving, the actual speed of the ball would be ninety miles per hour. The velocity of the car adds to the velocity of the thrown ball, which results in the ball moving at ninety miles per hour. This is about as fast as a major league baseball pitcher can throw a ball. However, when you postulate that the speed of light is the same in all inertial frames of reference, you are asserting that the frame of reference imparts no additional speed to the speed of light. Einstein termed this the principle of invariant light speed. The principle of invariant light speed has its roots in the Michelson–Morley experiment, performed in 1887. The Michelson–Morley experiment disproved the existence of "aether," an imaginary material that light was theorized to travel through. The experiment was repeated in 1902 and 1905, and the initial findings were confirmed with even greater accuracy. The nature of the experiment also suggested that the speed of light is constant in all inertial frames of reference.

I think it is important that the reader understand the historical context that led to Einstein's above postulates. Like many great scientists, Einstein based his postulates on the scientific work that others had done before him. There is a common misconception regarding Einstein. Some people believe that since he was a theoretical physicist, he worked in isolation, shutting himself off from the world of science. Nothing could be further from the truth. Einstein developed his theories in concert with the theoretical and experimental work of other scientists. This is not to suggest he was part of a "team," or that his work was not original, but rather to suggest he was well read and well versed in the scientific research of others. As elegant as the special theory of relativity is, he would have modified it or discarded it if it did not align with the theoretical and experimental data available at the time. In Einstein's own words, taken from his *Autobiographical Notes*, 1949, he stated, *"The insight fundamental for the special theory of relativity is this: The assumptions relativity and light speed invariance are compatible if relations of a new type ('Lorentz transformation') are postulated for the conversion of coordinates and times of events...The universal principle of the special theory of relativity is contained in the postulate: The laws of physics are invariant with respect to Lorentz transformations (for the transition from one inertial system to any other arbitrarily chosen inertial system). This is a restricting principle for natural laws...."* In other words, Einstein was well aware of the Lorentz transformation (discussed in appendix 5), as well as other scientific works. In particular, his position as a patent examiner required he read the references cited in the patent applications.

Today, scientists hold the special theory of relativity as the gold standard, having withstood over one hundred years of experimental verification. Scientists argue that other theories must meet similar standards to be scientifically valid. On a side note, this is the biggest criticism leveled against string theory. To date, we are unable experimentally to prove string theory's validity. Therefore, string theory remains an elegant mathematical formulation. We cannot definitively say string theory correctly describes reality, but that is another subject entirely.

Einstein's special theory of relativity gave us numerous new important insights into reality, among them the famous mass equivalence formula ($E = mc^2$) and the time dilation formula, which is provided in appendix 1, for those who want to delve deeper into the mathematics. I have also provided a sample calculation involving a ten-year trip on a spacecraft moving close to the speed of light. Can you guess how many years will have passed on Earth during the trip?

According to special relativity's time dilation, as a clock moves close to the speed of light, time slows down relative to a clock at rest. The implication is that if you were able to travel in a spaceship that was capable of approaching the speed of light, a one-year round trip journey as measured by you on a clock within the spaceship would be equivalent to approximately ten or more years of Earth time, depending on your exact velocity. In effect, when you return to Earth, you will have traveled to Earth's future. This is not science fiction. As I mentioned above, time dilation has been experimentally verified using particle accelerators. It is widely considered a science fact.

At this point, you may have a question: Why does a particle with a relatively short life have its life increased by accelerating it near the speed of light? In the next chapter, I will discuss an original theory, the existence equation conjecture, which demonstrates that time dilation has to do with a particle's energy. The greater the particle's kinetic energy, the longer it will exist. I introduced the existence equation conjecture to the scientific community in my book *Unraveling the Universe's Mysteries*, but more about that in the next chapter.

Einstein used the term "special" when describing his special theory of relativity because it only applied to inertial frames of reference, which are frames of reference moving at a constant velocity or at rest. It also did not incorporate the effects of gravity. Shortly after the publication of special relativity, Einstein began work to consider how he could integrate gravity and noninertial frames into the theory of relativity. The problem turned out to be monumental, even for Einstein. Starting in 1907, his initial thought experiment considered an observer in free fall. On the surface, this does not sound like it would be a difficult problem for Einstein, given his previous accomplishments. However, it required eight years of work, incorporating numerous false starts, before Einstein was ready to reveal his general theory of relativity.

Einstein's general theory of relativity (backward time travel)

In November 1915, Einstein presented his general theory of relativity to the Prussian Academy of Science in Berlin. The equations Einstein presented, now known as Einstein's field equations, describe how matter influences the geometry of space and time. In effect, Einstein's field equations predicted that matter or energy would cause spacetime to curve. This means that matter or energy has the ability to affect, even distort, space and time.

Einstein's field equations involve extremely difficult mathematics. They require significant depth in the mathematical disciplines of differential equations and geometry to understand them. Solutions to the equations are equally difficult. Even Einstein did not present exact solutions to the equations when he first introduced them. Instead, he presented approximations to the equations to explain his theory and its predictions. As a side note, in 1916, astrophysicist Karl Schwarzschild found the first nontrivial exact solution to Einstein's field equations. The solution is termed the Schwarzschild metric. Due to the mathematical complexity, we will only conceptually discuss the general theory of relativity. Most importantly, we will discuss its predictions and work our way to its implications regarding time travel.

Einstein's general theory of relativity provided a geometric interpretation of gravity, significantly differing from Newton's classical law of gravity. Instead of viewing gravity as an attractive force between two masses, general relativity views the masses as curving the space between them. This curvature of space causes the masses to come together, much as two marbles would fall together if placed at opposite sides of a pit in the ground. If you picture one marble on one side of a pit, and another marble on the opposite side of the same pit, when you release them, they fall to the bottom of the pit. Einstein's general theory of relativity predicts that matter curves space and time, which creates something like a pit in spacetime, which ultimately results in what we observe as gravity.

Is this geometric interpretation of gravity correct? Maybe! Does the answer surprise you? The reason for the "maybe" is that general relativity, which successfully deals with macro phenomena (i.e., large objects of our everyday world, including the orbits of planets), doesn't reconcile with the most successful theory dealing with quantum phenomena (i.e., the level of atoms and subatomic particles), quantum mechanics. The two most successful theories in science, general relativity and quantum mechanics, do not come together to give us a self-consistent view of gravity. Although there are ongoing attempts to develop a "quantum gravity" theory, which would combine both theories, no theory of quantum gravity has found wide acceptance in the scientific community. Nonetheless, general relativity is a successful theory and provides numerous insights and important predictions. Many of the predictions of general relativity have been scientifically verified. Two of the most important predictions for our study of time travel are (1) gravitational time dilation and (2) closed timelike curves.

Gravitational time dilation predicts that a clock in a strong gravitational field will run slower than a clock in a weak gravitational field. Therefore, a clock on the surface of Jupiter, a massive gas planet three hundred times larger than the Earth, resulting in a significantly stronger gravitational field, will run much slower than a clock on the surface of the Earth. This phenomenon was first verified on Earth, with clocks at different altitudes from the Earth's surface. Using atomic clocks, time dilation effects are detectable when the clocks differ in altitude by as little as one meter.

Gravitational time dilation also occurs in accelerating frames of reference (i.e., noninertial frames of reference). According to Einstein's general theory of relativity, an accelerated frame of reference produces an "inertial force," also termed a "pseudo force," that results in the same effect as a gravitational force in an inertial frame of reference. The equivalence of the inertial force in a noninertial frame of reference (i.e., an accelerating frame of reference) to a gravitational force in an inertial frame of reference (i.e., a frame of reference moving at a constant velocity) is termed the equivalence principle. The equivalence principle refers to the equivalence of "inertial mass" and "gravitational mass." Therefore, a blindfolded person in a rapidly ascending elevator would experience a force equivalent to an increase in gravity, as if standing on a planet more massive than Earth. The blindfolded person would not be able to determine if the force experienced is inertial or gravitational. This effect also holds true for time dilation. Time moves slower in a highly accelerated frame of reference in much the same way it would as if it were in a strong gravitational field. It is important to note, a frame of reference can accelerate in two fundamental ways. It can accelerate along a straight line, or it can accelerate by rotating. Formulas for time dilation due to gravitational and noninertial frames (i.e., accelerating frames) are in appendix 1, for those who would like to delve deeper into the mathematics.

Next, let us discuss closed timelike curves. What is a closed timelike curve? It is an exact solution to Einstein's general relativity equations demonstrating a particle's world line (i.e., the path the particle follows in four-dimensional spacetime) is "closed" (i.e., the particle returns to its starting point). Closed timelike curves theoretically suggest the possibility of backward time travel. The particle's world line is describable by four coordinates at each point along the world line, and when it closes on itself, the four coordinates at the start equal the four coordinates at the end. The particle, conceptually, went back to its past (i.e., the starting point). You can think of this like a horse racetrack. As the horse runs around the track, the horse eventually crosses the finish line, the starting point. If we allow the horse racetrack to represent a world line, then when the horse crosses the finish line, the horse has returned to its past (i.e., the starting point). In the mathematics of general relativity, the starting four coordinates, including the fourth dimensional coordinate that includes a time component, equal the four coordinates at the finish line.

The first person to discover a solution to Einstein's general relativity equations suggesting closed timelike curves (CTCs) was Austrian American logician, mathematician, and philosopher Kurt Gödel, in 1949. The solution was termed the Gödel metric. Since 1949, numerous other solutions containing CTCs have been found, such as the Tipler cylinder and traversable wormholes, both of which will be discussed in section 3. The numerous solutions to Einstein's general relativity equations suggest that time travel to the past is theoretically possible. However, the entire scientific community is not in complete agreement on this last point.

The largest issue that physicists have with backward time travel is causality violations (cause and effect), where the effect proceeds the cause. These violations of causality are termed "time travel paradoxes." Some physicists suggest that time travel paradoxes inhibit backward time travel, while other physicists argue that time travel paradoxes can be reconciled, and backward time travel is possible. There is no scientific consensus regarding the reality or practicality of time travel to the past. A wealth of time dilation data demonstrates forward time travel is both theoretically and experimentally possible. No equivalent body of scientific data demonstrates backward time travel is an achievable reality. Some physicists even argue that further refinements to general relativity to reconcile it with quantum mechanics may rule out the closed timelike curve solutions to general relativity. As you can likely imagine, this has led to heated discussions among some of the greatest physicists in the world. We will discuss time travel paradoxes, along with the current scientific discussions, in a later chapter.

Conclusions from Chapter 3

- Einstein's special theory of relativity provides a strong theoretical foundation for forward time travel, which is termed "time dilation."

- There is a wealth of scientific data proving time dilation is real and can occur when a frame of reference accelerates near the speed of light, or when a frame of reference is in a strong gravitational field.

- Even though there is general agreement regarding time dilation, no one has built a machine that enables a human to experience significant time dilation. It is true, however, that people traveling at high speeds, like astronauts, experience some time dilation. To date, the amount of time dilation experienced by any humans is only a small fraction of a second, and not noticeable to the humans involved.

- Particle accelerators, such as the Large Hadron Collider, are able to accelerate subatomic particles near the speed of light, and time dilation is a measurable effect.

- Einstein's general theory of relativity predicts gravitational time dilation. The scientific community generally agrees time dilation occurs in strong gravitational fields.

- Some solutions to Einstein's equation of general relativity result in closed timelike curves, which theoretically suggest backward time travel.

- The scientific community is not in agreement regarding the practicality and reality of backward time travel. In fact, the entire subject of backward time travel is contentious. We will discuss this, as well, in a later chapter.

This chapter delineates the traditional science regarding time travel. As is evident from reading it, the traditional science relies on Einstein's theories of special and general relativity as a foundation for time travel. However, the traditional science does not give us the fundamental mechanism regarding time travel. For example, Einstein's theories predict time dilation (forward time travel), but they do not explain the fundamental cause. This significantly bothered me and led me to do theoretical research using Einstein's theories as expressed geometrically in Minkowski spacetime. The result is a new theory that explains time dilation on the fundamental basis of energy. I term this new theory the existence equation conjecture. I put it forward in my book *Unraveling the Universe's Mysteries*. I will also discuss it here. If this new theory proves valid, then we are one step closer to understanding how to build a time machine, and even explain the accelerated expansion of the universe. Are you interested in looking behind the curtain and understanding why time dilates? The answer is one page away.

CHAPTER 4

The Existence Equation Conjecture

In 1905, Einstein originally published his special theory of relativity in algebraic form. For example, Einstein's famous mass-equivalence formula ($E = mc^2$) is an algebraic equation that states energy (E) is equal to the rest mass (m) multiplied by the speed of light (c) squared (i.e., the speed of light multiplied by itself). It is the most famous equation in the world.

In 1907, Einstein's former professor, Polish mathematician Hermann Minkowski (1864–1909), reformulated special relativity geometrically in four-dimensional spacetime, known as Minkowski spacetime. At the time, the twenty-eight-year-old Einstein did not make much of this new mathematical treatment of special relativity, considering it something of a curiosity. Over the next eight years, that was going to change for Einstein and the world. The geometric treatment of special relativity became an important building block for Einstein's general theory of relativity.

The energy required to exist

The title of this chapter is "The Existence Equation Conjecture." Let me emphasize the word "conjecture," which means that the existence equation is my opinion and requires peer review. I believe the existence equation, including my interpretation of it, is correct. However, I will continue to label it a conjecture until it has been rigorously peer reviewed.

What is the existence equation conjecture? To understand the existence equation conjecture, let us start with something that most of us are familiar with, namely, kinetic energy. Kinetic energy is the energy associated with an object's motion, and it is directly proportional to the velocity squared (i.e., the velocity multiplied by itself) and the mass of the object. For example, if you throw a baseball at a window, it will likely break the glass. This occurs because the baseball has kinetic energy. If you throw a small cotton ball, about the size of a marble, against a window, it will likely just bounce off without breaking the glass. The reason is the baseball's kinetic energy is higher due to its greater mass, even if you throw the cotton ball at the same velocity.

When a mass moves in any of the three spatial dimensions, it has kinetic energy. Everyone, including high school science students, agrees with this statement. However, an important question is seldom asked. What is the kinetic energy associated with a mass's movement in the fourth dimension of Minkowski spacetime? In my book *Unraveling the Universe's Mysteries*, I addressed this question. To address the question, I had to derive the relativistic kinetic energy associated with a mass's movement in the fourth dimension of Minkowski spacetime. Since this is presented in appendix 2, I will not go into further detail here regarding Minkowski spacetime. However, it is important for you to know that Minkowski's spacetime coordinates inspired me to think about a mass's movement (i.e., velocity) in the fourth dimension. Using a mass's velocity in the fourth dimension of Minkowski spacetime, along with Einstein's relativistic formula for kinetic energy, enabled me to derive the existence equation conjecture. For those who wish to delve deeper into the mathematics and experimental verification of the existence equation conjecture, please consult appendices 2 and 3.

Here is the existence equation conjecture:

$$KE_{X4} = -.3mc^2$$

Where KE_{X4} is the kinetic energy associated with an object's movement in the fourth dimension of Minkowski spacetime, m is the rest mass of an object, and c is the speed of light in a vacuum.

What does this all mean? Phrases like "Minkowski spacetime," "fourth dimension," "relativistic formula for kinetic energy," and the mathematics and experimental verification in appendices 2 and 3 tend to confuse most people. This is a specialized niche and requires an understanding of special relativity, advanced mathematics, and Minkowski spacetime. For these reasons, I am not going to labor here on the derivation and the experimental verification. Those readers who are interested may consult appendices 2 and 3. However, I will provide a conceptual interpretation. With the caveat that the interpretation is speculative, here is my interpretation: I interpret the equation to imply that a mass requires energy to move in the fourth dimension of Minkowski spacetime. The implication of requiring energy comes from the observation that the kinetic energy is negative. Although Einstein never called the fourth dimension time, it includes a time component, and I interpret the mass's movement in the fourth dimension to relate to its existence. To explain this last point, think of this simple example. If you are observing an object held in your hand, you and the object are moving in time at the same rate. You can affirm the object exists. If the object, for some unknown reason, stopped moving in time, it would disappear from your hand. This illustrates movement in time and existence is intimately related.

The equation is dimensionally correct (expressible in units of energy), but highly unusual from two standpoints:

1. The kinetic energy is negative. For masses moving in the typical three-dimensional space, the kinetic energy is always positive. This is independent of the size of the mass. For example, it is true of bowling balls, baseballs, golf balls, atoms, and subatomic particles. The only particles known to science to exhibit negative kinetic energy are virtual particles. These particles pop in and out of existence in a vacuum, and they are responsible for numerous physical effects, such as the Casimir effect. For those of you who are not familiar with the Casimir effect, here is a brief explanation. Two closely spaced electrically neutral plates within a vacuum experience a force, known as the Casimir-Polder force, which pushes the plates together. This is caused by fewer virtual particles being able to form between the closely spaced plates relative to the significantly larger number of virtual particles that form on the outside surface of the plates. Please see figure 1.

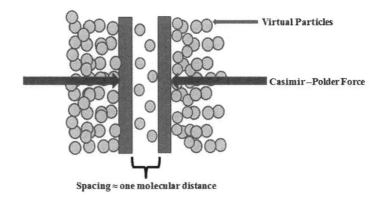

Figure 1: Casimir Effect

The Casimir effect demonstrates that vacuums contain energy and can give rise to virtual particles. Although counterintuitive, a laundry list of effects demonstrates that virtual particles are real, and that vacuums contain energy.

2. The amount of negative kinetic energy suggested by the equation is enormous. For example, even something with a small mass, like a golf ball, would be equivalent to the energy of an atomic bomb, only the energy is negative. Unfortunately, though the scientific community acknowledges negative energy exists, science knows very little about it. Negative energy is typically associated with the hypothetical existence of negative mass. For this reason, a common misconception about antimatter is that it is negative energy. Antimatter is not negative energy. It is still positive energy. Some physicists are exploring the Casimir effect to determine if it can produce negative energy.

To summarize, the existence equation conjecture is the amount of energy a mass (*m*) requires to move in the fourth dimension of Minkowski spacetime, which relates to its existence. In appendix

3, I provide an example. Using the existence equation conjecture, I determined how much relativistic kinetic energy a muon (i.e., a negatively charged subatomic particle about two hundred times heavier than an electron) would require to extend its life span by 29.3 (i.e., exist 29.3 times longer than it would at rest), and compared that to particle acceleration data. The predicted life span using the existence equation conjecture was within 2% of the energy associated with the particle acceleration data.

The above result is important from two standpoints:

1. It gives credence to the existence equation conjecture's ability to predict the life span (i.e., time dilation) of a mass as a function of its kinetic energy. For example, if we want the muon's life span to increase by 50 times its rest mass life, then we need to multiply ($.3mc^2$) by 50 and solve for the relativistic kinetic energy. This in turn will enable us to determine the velocity the particle must have to reach a life span (i.e., time dilation) 50 times greater than its rest mass. Notice, we are adding positive kinetic energy to balance the negative kinetic energy required for existence.

2. It gives credence to the interpretation of the existence equation conjecture, namely, that it equates to the amount of energy a particle requires (i.e., implied by the negative sign) to exist (i.e., move in the fourth dimension of Minkowski spacetime).

Therefore, the fundamental cause of time dilation due to motion is the addition of positive kinetic energy. When we add positive kinetic energy to a mass, we offset the negative kinetic energy it requires to move in the fourth dimension of Minkowski spacetime. This in turn increases the particle's existence.

Does the existence equation conjecture also predict the energy required for gravitational time dilation? In theory, it should, but validating that is much more difficult. First, the Earth is a noninertial frame of reference. Stating the Earth is a noninertial frame of reference means it is a frame of reference undergoing acceleration, which is due to its rotation about its own axis as well as its orbit around the Sun. The noninertial aspects of the Earth as a frame of reference add to the gravitational time dilation effect. Second, the effects of the Sun's gravitational effects also would be a factor in the resulting time dilation. To get around the Earth's rotation, plus the effects of the Sun's gravity, a recent experiment on gravitational time dilation used two clocks, both located in the same noninertial frame of the Earth (C. W. Chou, D. B. Hume, T. Rosenband, and D. J. Wineland, "Optical Clocks and Relativity," *Science*, vol. 329, no. 5999, September 24, 2010, pp. 1630–1633). The experiment confirmed that the clock closer to the Earth has a longer second than a clock about twelve inches higher. By only comparing differences in the length of a second between the two clocks, Chou and his colleagues were able to remove the effect of the Earth's rotation and the influence of the Sun's gravitational field. However, the measured differences are extremely small. For example, based on the referenced experiment, a person living twelve inches lower than another person would only add about ninety billionths of a second to a seventy-nine-year life span. We do not encounter similar errors when we accelerate a particle to near the speed of light using a particle accelerator. The acceleration and associated kinetic energy render the time dilation associated with the Earth's gravity and rotation, or the Sun's gravitational influence, insignificant. To correlate the existence equation conjecture to the Chou experiment would require including the Earth's rotation, plus the Sun's gravitational influence. This calculation is difficult, involving the equations of general relativity, and is beyond the scope of this book. My point in presenting the above information is to provide an insight into the difficulties of calculating the total gravitational energy. However, since the existence equation conjecture accurately predicts the kinetic energy associated with muon's time dilation in the Bailey experiment, as discussed in chapter 1 and in appendix 3, my judgment is that it would predict the gravitational time dilation effects as well, if someone were able to perform the appropriate calculations.

Positive energy—time traveling to the future

The existence equation conjecture accurately predicts the positive energy required for time dilation associated with kinetic energy of subatomic particles. It gives us a clear picture of the role positive kinetic energy plays in time dilation, which is forward time travel. This begs another important question: Does the existence equation conjecture imply that adding negative energy to a mass will cause it to travel backward in time?

Negative energy—time traveling to the past

Today's science knows precious little about negative energy. The best example we have of creating negative energy in the laboratory is the Casimir effect, which we briefly discussed previously, but will now discuss in detail. Let us start by discussing the energy associated with a vacuum. Vacuums contain energy. One simple experiment to prove this is to take two electrically neutral metal plates and space them closely together in a vacuum. They will be attracted to each other (i.e., the Casimir effect). At approximately 10 nm (i.e., 1/100,000 meters) separation, the plates experience an attraction force of about one atmosphere (i.e., typically, the pressure we feel at sea level on Earth). What is causing this force?

The energy in a vacuum is termed "vacuum energy." Surprisingly, it appears to obey the laws of quantum mechanics. For example, the energy will statistically vary within the vacuum. When the vacuum energy statistically concentrates, it gives rise to virtual particles, which is termed a "quantum fluctuation." When the metal plates are spaced closely, relatively few virtual particles can form between the plates. A much larger population of virtual particles can form around the plates. This larger population of particles exerts a force on the outside of the plates. This force is the Casimir-Polder force, and it pushes the plates together. However, another strange physical phenomenon is also occurring between the closely spaced plates. In quantum mechanics, every particle has a "zero-point energy." Even a vacuum is said to have a zero-point energy. The zero-point energy, or the "ground state," is the lowest energy level that a particle or a vacuum may have. By reducing the space between the plates, some physicists believe we are reducing the normal zero point energy of the vacuum between the plates. When this occurs, those physicists argue the vacuum energy between the plates is negative energy (i.e., below the zero-point energy).

The scientific community is not in complete consensus regarding the properties or even the existence of negative energy. Physicists are able to mathematically model negative energy and use those models to make predictions regarding the theoretical behavior of negative energy. While the mathematical models do not prove the existence of negative energy, it is instructive to consider their predictions, and their implications to time travel. Here are the salient features of negative energy based on the mathematical modeling:

- Negative energy implies the existence of negative mass. This, of course, begs a question. What is negative mass? Negative mass is a hypothetical concept in theoretical physics. Anglo-Austrian mathematician and cosmologist Hermann Bondi suggested its existence in 1957. If it exists, it is the negative counterpart of normal (i.e., positive) mass and exhibits unusual properties. For example, normal masses exhibit attractive forces, known as gravitational attraction. Negative masses would exhibit repulsive forces. However, be careful not to equate negative mass with antimatter. The vast majority of the scientific community holds that antimatter is still positive mass. Based on this consensus, they predict antimatter would exhibit the same properties as positive mass. For example, two antimatter particles would exert an attractive force on each other, not a repulsive force. The implications of negative mass on time travel are ambiguous, since the existence of negative mass itself is ambiguous.

- Several in the scientific community suggest that a negative energy vacuum would allow light to travel faster than a normal positive energy vacuum. If this theory proves to be correct, it could have major implications for time travel. For example, there is speculation that this property may allow people to travel faster than the speed of light in a negative-energy vacuum bubble. Previously, we have discussed that as a mass approaches the speed of light, time dilates (i.e., time slows down for the mass). If the mass exceeds the speed of light, the implication is that it can travel into the past. We will discuss this further in the next chapter.

- Stephen Hawking and other physicists suggest that negative energy is required to stabilize a "traversable wormhole," an entity that would allow a person, object, or information to travel between two points in time or space. Wormholes are a hypothetical shortcut between two points in time or two points in space. There are solutions to Einstein's general equations of relativity suggesting the theoretical existence of wormholes. However, we have no observational evidence that they exist in reality. We will discuss traversable wormholes in the next chapter.

Until we can find a way to produce negative energy and apply it experimentally to determine its effect on time, we can only speculate. The existence equation conjecture, though, gives us a fundamental basis to make those speculations.

Conclusions from Chapter 4

- Einstein's equations of special and general relativity predict time dilation due to a mass's movement, and/or a mass's position in a gravitational field. However, relativity does not give us the fundamental mechanism causing time dilation.

- The existence equation conjecture does provide the fundamental mechanism for time dilation related to a mass's movement.

- The existence equation conjecture is a speculative theory. That is why the word "conjecture" is used.

- With the above caveat, I interpret the equation to imply that a mass requires energy to move in Minkowski's fourth dimension of spacetime, which I equate with the mass's existence.

- The existence equation conjecture may apply to gravitational time dilation, but proving this is difficult and requires more data.

- An interesting speculation regarding the existence equation conjecture relates to negative energy. Does the existence equation conjecture imply adding negative energy to a mass will cause it to travel backward in time? I speculate it does, but in the absence of experimental evidence and with the current level of scientific knowledge about negative energy, this is highly speculative.

This chapter completes the major theories that underpin the science of time travel. It is one thing to have a theory. It is another thing to apply it. You may wonder: What does science propose to time travel? The next section addresses building time machines. Are you ready for a temporal shock? Please read on.

SECTION III

Building a Time Machine

"Once confined to fantasy and science fiction, time travel is now simply an engineering problem."

—**Michio Kaku**
***Wired* magazine, August 2003**

"I myself believe that there will one day be time travel because when we find that something isn't forbidden by the overarching laws of physics we usually eventually find a technological way of doing it."

—David Deutsch (1953 –)
British physicist at the University of Oxford

CHAPTER 5

What Methods Does Science Propose for Time Traveling?

Various scientists have proposed numerous methods to time travel. We will examine some of the most popular methods in this chapter. Sadly, I must report that they all share a common issue. With the exception of time dilation experiments (i.e., forward time travel), the proposed methods are beyond today's science, especially for backward time travel. A small amount of data regarding backward time travel does exist, which we discussed in chapter 1 (i.e., the section titled "Twisting the arrow of time"). However, the amount of experimental data regarding backward time travel (i.e., closed timelike curves) is scant compared to forward time travel (i.e., time dilation).

To date, no method of time travel involving humans exists. We currently do not have a working time machine. However, exploring the methods that science proposes for time travel is still useful. The proposed time travel methods enable a greater understanding regarding time travel science and provide insight into the engineering challenges associated with building a practical time travel machine.

All the methods discussed in this chapter have their theoretical foundation in either Einstein's special or general theory of relativity. We will now examine the most popular proposed methods for time travel.

Proposed time travel methods

1. Faster-than-light time travel—Based on Einstein's special theory of relativity, if we are able to move information or matter from one point to another faster than the speed of light, there would be some inertial frame of reference (i.e., a frame of reference moving at a constant velocity) in which the signal or object is moving backward in time. Let us understand why this is the case.

    Consider sending a signal from one location to another. The first event is sending the signal. The second event is receiving the signal. As long as the signal travels at or below the speed of light, according to the "relativity of simultaneity," the first event will always precede the second event in all inertial frames of reference. Although this squares with our everyday observation of reality, that cause precedes effect, you may have a question. What is the relativity of simultaneity?

The relativity of simultaneity is a concept introduced by Einstein in the special theory of relativity. The simultaneity of an event is not an absolute to all observers, but depends on the observer's frame of reference. For example, if one observer is midway on a train car, and a second observer is at rest on the platform at the train station, they will see the simultaneity of an event differently. As the two observers pass, assume the observer in the train takes a picture using a flashbulb. From the viewpoint of the observer within the train, the light reaches both the front and rear of the train car at the same time. However, the observer on the platform sees a different situation. From the observer on the platform's viewpoint, first the flashbulb goes off, and then the light reaches the back of the train car, since it was moving toward the fixed observer on the platform. Lastly, the observer sees the light reach the front of the train car, since it was moving away from the observer. The effect is more pronounced as the speed of the train approaches the speed of light.

Based on the relativity of simultaneity, if a signal propagates faster than the speed of light, there would always be some frames of reference where the signal arrives before it was sent. To illustrate this, let us go back to the above example and assume the train is traveling close to the speed of light. The observer is now closer to the end of the train car when the flashbulb flashes. Let us also assume the light exceeds the speed of light in a vacuum. For example, we could assume the interior of the train car contains a negative energy vacuum, which some in the scientific community believe would allow light to travel faster than it would in a normal positive vacuum. Given these two inertial frames of reference, the train moving close to the speed of light, and the observer situated closer to the rear of the train car when the flashbulb goes off, it would appear that the light reached the end of the train car prior to the light from the flashbulb reaching the observer on the platform. Why is this? Please refer to figure 2 below. The light inside the train instantaneously reaches the back of the train car (light path one), and then travels a short distance in the inertial frame of the observer (light path two), who records the event. This is witnessed ahead of the light reaching the observer from the source (light path three), since now the observer is farther away from the source. Therefore, the observer first witnesses the light reach the back of the train, and then observes the light from the source. From the viewpoint of the observer at the station, the effect preceded the cause. If the light within the train did not travel faster than the speed of light in a vacuum, the light paths one plus two would take longer to reach the observer than light path three. Thus, the effect of reverse causality would be lost.

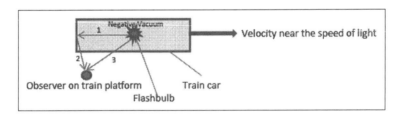

Figure 2: Train Car Example of Faster-Than-Light Travel
Affecting Causality

From inside the train car, nothing changes for the observer seated midway in the car. The faster-than-light signals reach the front and back at the same time. In summary, the observer on the platform witnesses reverse causality. The light signal reaches the back of the train car before the light from the flashbulb reaches the observer on the platform. This thought experiment, illustrating reverse causality, suggests the observer on the platform witnesses an event taking place in the past (i.e., light reaching the end of the train car), since the flashbulb light at the source will reach the observer on the platform later (i.e., the future).

2. Traversable wormholes—In this section, we will discuss traversable wormholes in significant detail. Let us begin our discussion by understanding the scientific meaning of a "wormhole." There are valid solutions to Einstein's equations of general relativity that suggest it is possible to have a "shortcut" through spacetime. To picture this, consider a piece of paper with a dot at opposite corners. In Euclidian geometry, normally taught in high school, we learn that the shortest distance between the two points is a straight line. However, valid solutions to Einstein's general relativity equations suggest that the two points on the paper are connectable by an even shorter path, a wormhole. To visualize this, simply fold the opposite corners of the paper with the dots, such that the dots touch. You have created a representation of a wormhole. You have manipulated the space between the dots by folding the paper to allow them to touch.

Unfortunately, there is no scientific evidence that wormholes exist in reality. However, the strong theoretical foundation suggesting wormholes (i.e., valid solutions to Einstein's equations of general relativity) makes their potential existence impossible to ignore.

The first type of wormhole solution to Einstein's equations of general relativity was the Schwarzschild wormhole, developed by German physicist Karl Schwarzschild (1873–1916). Unfortunately, although the Schwarzschild mathematical solution was valid, it resulted in an unstable black hole. The unstable nature of the Schwarzschild wormhole suggested it would collapse on itself. It also suggested that the wormhole would only allow passage in one direction. This brought to light an important new concept. Faced with the unstable nature of Schwarzschild wormholes, American theoretical physicist Kip Thorne and his graduate student Mike Morris demonstrated a general relativity "traversable wormhole" in a 1988 paper. In this mathematical context, a traversable wormhole would be both stable and allow information, objects, and even humans to pass through in either direction and remain stable (i.e., would not collapse on itself). As is often the case in science, one discovery leads to another. Numerous other wormhole solutions to the equations of general relativity began to surface, including one in 1989 by mathematician Matt Visser that did not require negative energy to stabilize it.

As discussed above, traversable wormholes may require negative energy to sustain them. Several prominent physicists, including Kip Thorne and British theoretical physicist/cosmologist Stephen Hawking, believe the Casimir effect proves negative energy densities are possible in nature. Currently, physicists are using the Casimir effect, which we discussed in chapter 4, in an effort to create negative energy. Obviously, if successful, the amounts of negative energy will likely be small. Because of the amount of negative energy that may result, I suspect the first wormholes developed will be at the quantum level (i.e., the level of atoms and subatomic particles).

We have merely scratched the surface regarding the science of wormholes, but we did accomplish one important objective. We have described how a traversable wormhole would allow spacetime travel via shortcuts in spacetime. This means we could connect two points in time or two points in space via a traversable wormhole. However, there is a hitch regarding time travel to the past. According to the theory of relativity, we cannot go back to a time before the wormhole existed. This means that if we discover how to make a traversable wormhole today, a year from now we can go back to today.

You may wonder why a wormhole constructed today would not allow us to go back to yesterday. To understand this conundrum, we need to understand just how a wormhole works as a time machine. Here is one scenario. Imagine you are able to accelerate one end of a wormhole to a significant fraction of the speed of light. Perhaps you could use a high-energy ring laser (i.e., a laser than rotates in a circle). As you twist the space, you create the "mouth" of the wormhole, something like a tunnel. After you enter the mouth of the wormhole, you are now somewhere in the wormhole's "throat." A "tunnel" is a good analogy to what is occurring. Now imagine you are able to take the other entrance of the tunnel, which is at rest and called the "fixed end," and bring it back close to the origin. Time dilation causes the mouth to age less than the fixed end. A clock at the mouth of the wormhole, where spacetime accelerates near the speed of light, will move slower than a clock at the fixed end.

Given the above understanding of how a wormhole acts as a time machine, let us address why it is only possible to go back to the time of the wormhole's construction. Imagine you have two synchronized clocks. If you place one clock at the mouth, and you place the other clock at the fixed end, they will initially read exactly the same time, for example, the year 2013. However, the clock at the mouth, influenced by the twisted space, is going to experience time dilation, and therefore move slower than the clock at the fixed end. Let us consider the case where the clock at mouth of the wormhole moves, based on the rate of twisting spacetime, one thousand times slower than the clock at the fixed end. In one hundred years, the clock at the fixed end, which experiences no time dilation, will read 2113. The clock at the mouth will still read 2013; only one tenth of one year will have passed due to time dilation at the mouth of the wormhole. From the fixed end, where no time dilation is occurring (i.e., the clock reads 2113), you can walk back to the mouth of the wormhole, where the clock still reads 2013. You will have walked one hundred years into the past. Notice, though, you cannot go back beyond the time of the traversable wormhole's construction.

3. Tipler cylinder time travel—The Tipler cylinder is a cylinder of dense matter and infinite length. Historically, Dutch mathematician Willem Jacob van Stockum (1910–1944) found Tipler cylinder solutions to Einstein's equations of general relativity in 1924. Hungarian mathematician/physicist Cornel Lanczos (1893–1974) found similar Tipler cylinder solutions in 1936. Unfortunately, neither Stockum nor Lanczos made any observations that their solutions implied closed time-like curves (i.e., time travel to the past). In 1974, American mathematical physicist/cosmologist Frank Tipler's analysis of the above solutions uncovered that a massive cylinder of infinite length spinning at high speed around its long axis could enable time travel. Essentially, if you walk around the cylinder in a spiral path in one direction, you can move back in time, and if you walk in the opposite direction, you can move forward in time. This solution to Einstein's equations of general relativity is known as the Tipler cylinder. The Tipler cylinder is not a practical time machine, since it needs to be infinitely long. Tipler suggests that a finite cylinder may accomplish the same effect if its speed of rotation increases significantly. However, this remains only a speculation, without proof. No Tipler cylinder time machine exists to date.

4. Alcubierre drive time travel—This is based on a solution to Einstein's general relativity equations by Mexican theoretical physicist Miguel Alcubierre. Dr. Alcubierre published a 1994 paper, "The Warp Drive: Hyper-Fast Travel Within General Relativity," in the science journal *Classical and Quantum Gravity*.

The Alcubierre drive appears to allow a spaceship to travel faster than light, but it requires the existence of negative mass to make the Alcubierre drive work. In principle, the drive works by contracting the space in front of the spaceship and expanding the space behind the spaceship faster than the speed of light. In this fashion, the spaceship rides like a surfer on a wave. As the space behind the spaceship expands faster than the speed of light, the spaceship appears to move faster than the speed of light. However, it does not. Only the space behind the ship is expanding faster than the speed of light. In this way, Dr. Alcubierre avoids violating the laws of special relativity, namely, that no mass can exceed the speed of light.

There is no law in physics that prohibits space from expanding faster than the speed of light. From this viewpoint, the Alcubierre drive has merit. The Alcubierre drive is a mathematically valid solution to Einstein's field equations. However, requiring negative mass as part of the mechanism for the Alcubierre drive makes the theory highly speculative and, once again, beyond the reach of today's science. As a side note, Dr. Alcubierre got this idea by watching *Star Trek* and its use of the warp drive.

5. Particle accelerators—The Large Hadron Collider (LCH) is an example of a real time machine. It was built by the European Organization for Nuclear Research (CERN) from 1998 to 2008,

and it is the world's largest and highest-energy particle accelerator. As previously discussed, building the LHC required the collaboration of over ten thousand scientists and engineers from over one hundred countries, as well as hundreds of universities and laboratories. It is one of the great engineering milestones of humankind.

The purpose of the LHC is fourfold:

1. Test the predictions of different theories of particle physics.

2. Test the predictions of different theories of high-energy physics.

3. Prove or disprove the existence of the theorized Higgs boson, often referred to as the "God particle," which allows other particles to acquire mass.

4. Prove or disprove the existence of the large number of particles predicted by the supersymmetric theories, most notably the WIMP particle associated with dark matter. Supersymmetric theories are modifications to the standard model of particle physics. (As a side note, experimental work done in this area using the LHC has not verified the supersymmetric theories, which has led to the CERN scientists suggesting the supersymmetric theories be abandoned.)

How does the LHC work as a time machine? The LHC has the capability to accelerate subatomic particles to a significant fraction of the speed of light. As a result, the particle experiences time dilation, or forward time travel. The effects of time dilation for an unstable particle, like a muon, which would normally have a rest life of about 10^{-6} seconds (i.e., 1 divided by 1 with six zeros after it) before decaying, can be increased a factor of ten or more.

6. Black hole time travel—What is a black hole? A black hole is a point in space where gravity pulls so much that not even light can escape. We cannot see black holes, but we can infer their existence by how they influence stars around them.

There are numerous types of black holes. Some are small, about the size of an atom. Yet, they can have a mass equal to a mountain. Some are supermassive, like the black hole theorized to exist at the center of our galaxy, a mere twenty-six thousand light-years from us. It is the single heaviest object in our galaxy. In between the atom-size black holes and the supermassive black holes are the "stellar" black holes. They are roughly up to twenty times the mass of our sun.

You may wonder: How do black holes form? Physicists think that the atom-size black holes formed during the early stages of the big bang, and that the supermassive black holes formed when the galaxies formed. Physicists also think the stellar black holes form when a star dies and collapses on itself.

What makes a black hole interesting from the standpoint of time travel is that the gravitational attraction is so great that time dilation due to gravity (as predicted by Einstein's general theory of relativity) would be enormous. In fact, a supermassive black hole, like the one at the center of our galaxy, would slow down time far more than anything else in the galaxy would. This makes a black hole a natural type of time machine.

You may worry that a black hole may swallow the Earth. However, I have good news for you. Black holes do not move around, and there are none close to the Earth. In short, we do not have to worry about being swallowed by a black hole.

Is there any practical way to use a black hole as a time machine? The answer is no, not via today's science. The scientists at CERN using the Large Hadron Collider are attempting to make small black holes. Perhaps, in time, they will succeed, and we will be able to use its properties as a time machine. This, however, is speculation.

7. Mallett spacetime twisting by light (STL)—Dr. Ronald Mallett is an American theoretical physicist and the author of *Time Traveler: A Scientist's Personal Mission to Make Time Travel a Reality* (2007). Dr. Mallett is a full professor at the University of Connecticut, where he has taught physics since 1975.

Dr. Mallett is attempting to twist spacetime using a ring laser (i.e., a laser that rotates in a circle) by passing it through a through a photonic crystal (i.e., a crystal that only allows photons of a specific wavelength to pass through it). The concept behind STL is that by twisting space via the laser, closed timelike curves will result (i.e., time will also be twisted). In this way, Dr. Mallett hopes to observe a violation of causality when a neutron is passed through the twisted spacetime. Dr. Mallett also believes he will be able to send communication by sending subatomic particles that have spin up and spin down. Note, the spin of a subatomic particle is part of the particle's quantum description. As a simple example, we can consider spin up equal to 1 and spin down equal to 0. Using this technique, Dr. Mallett can send a binary code, similar to the binary codes used in computing.

Few scientists openly discuss their work on time machines. They fear ridicule. In this regard, Dr. Mallett is a pioneer. When Dr. Mallett was ten years old, his father died at age thirty-three from a heart attack. Dr. Mallett has shared that his initial drive to invent a time machine was to go back in time and visit with his father. Unfortunately, the science of time travel only allows a person to go back in time to the point when the time machine is first turned on. Dr. Mallett acknowledges this, but continues his quest.

Dr. Mallett's concept of twisting space is close to the concept of creating a wormhole, as discussed above. Dr. Mallett is using laser light as means of creating the mouth of the wormhole. In a publication (R. L. Mallett, "The Gravitational Field of a Circulating Light Beam," *Foundations of Physics* 33, 1307–2003), Dr. Mallett argued that with sufficient energies, the circulating light beam might produce closed timelike lines (i.e., time travel to the past).

Is Dr. Mallett's theoretical foundation solid? According to physicists Dr. Olum and Dr. Everett, it is fatally flawed. In a paper published in 2005 (Ken D. Olum and Allen Everett, 2005, "Can a Circulating Light Beam Produce a Time Machine?", *Foundations of Physics Letters* 18 (4): 379–385), they argue three points:

1. Dr. Mallett's analysis contains unusual spacetime (i.e., mathematical) issues, even when the power to the machine is off.

2. The energy required to twist spacetime would need to be much greater than lasers available to today's science.

3. They note a theorem proven by Stephen Hawking (chronology protection conjecture—1992), namely,

it is impossible to create closed timelike curves in a finite region without using negative energy.

Although Dr. Mallett did not address their criticism in a formal publication, he did argue in his book, *Time Traveler*, that he was forced to simplify the analysis due to difficulties in modeling the photonic crystal. This, however, is far from a complete response.

Who is right? In the physical sciences, we are judged by the weakest link in our theories. If I use this criterion, I would say the argument favors Dr. Mallett, since the chronology protection conjecture, which we will discuss in the next chapter, has come under serious criticism, and it is not clear that it presents a valid challenge. Nonetheless, Dr. Olum and Dr. Everett are highly regarded physicists. Therefore, at this point, it is hard to know who is right, and right about what. Perhaps the mathematical analysis is flawed, and the approach published by Dr. Mallett requires more energy than is available via today's technology. However, we are witnessing a significant event in science. A respected physicist, Dr. Mallett, is openly publishing his work on building a backward time travel machine. Other respected physicists, Dr. Olum and Dr. Everett, are entering into a scientific debate regarding Dr. Mallett's theoretical basis. From my point of view, this is how it should be in science. The debate is healthy. As a theoretical physicist, I know that the debate will end only when either:

1. The Mallett time machine works, or

2. The Mallett time machine enters the rubbish pile of scientific failures, along with astronomer Ptolemy's Earth-centered model of the solar system and the flat Earth theories.

Are there any real time machines?

There are no existing time machines capable of sending humans forward or backward in time. The closest we have come to time travel is using particle accelerators to cause subatomic particles to experience time dilation (i.e., forward time travel). There is a significant amount of time dilation data available. Particle accelerators succeed in achieving time dilation by accelerating subatomic particles close to the speed of light. Unfortunately, though, backward time travel has no similar body of experimental data. The major problems with creating backward time travel appear to fall into three categories:

1. Backward time travel appears to require negative energy, based on arguments made by American theoretical physicist Kip Thorne and British theoretical physicist/cosmologist Stephen Hawking. Many in the scientific community acknowledge that negative energy likely exists, and point to the Casimir effect, discussed previously, as an example in nature. However, today's science is unable to harness negative energy in any meaningful way to make a time machine.

2. Many in the scientific community, like physicists Dr. Olum and Dr. Everett, believe the amount of energy required to twist space sufficiently for spacetime manipulation and enable Dr. Mallett's time machine to work is enormous. Conceptually, we may be talking about the amount of energy provided by a star, similar to our own sun. Harnessing this level of energy is far beyond today's science. Science's best efforts to study high-energy physics has to date been confined to particle accelerators, such as the Large Hadron Collider. There is no experimental evidence that Dr. Mallett has succeeded in manipulating spacetime.

3. Many in the scientific community are concerned with causality violations, especially regarding backward time travel. However, as we learned in the section titled "Twisting the arrow of time," there can also be causality violations regarding forward time travel. The causality violations are generally termed "time travel paradoxes," which we will discuss in detail in the next chapter.

Having made the above points, I think it is important to point out that some physicists believe subatomic antimatter particles travel in the opposite direction in time (i.e., backward in time) versus their matter counterparts. For example, some physicists assert that positrons, the antimatter equivalent of electrons, travel backward in time, while electrons travel forward in time. In solid-state physics, if we consider a current flowing in a semiconductor, electrons in a semiconductor move as a current in one direction, while the "holes" (i.e., the position the electron occupied in the semiconductor, which becomes vacant when the electron moves as a current) move in the opposite direction. Physicists differ on whether the "holes" represent positrons (i.e., actual physical antimatter particles). I mention this for completeness. There is no scientific consensus that antimatter travels backward in time.

Where does this leave us? I think this question deserves a complete answer. In my opinion, we are in about the same place space travel was at the beginning of the twentieth century. At the beginning of the twentieth century, all we knew about space travel came from science fiction. We knew that birds could fly, and this observation provided hope that human air flight would eventually be possible. However, at this point we could only fly using balloons, which was a long way from controlled air flight. We knew about projectiles, such as cannonballs and simple rockets, and this provided hope that one day humankind would be able to travel into space. However, at the beginning of the twentieth century we were still three years away from building the first successful airplane. The first successful airplane did not come from a well-respected theory or formal scientific investigation. Most early attempts at air flight tended to focus on building powerful engines, or they attempted to imitate birds. The early attempts at air flight were dismal failures. The first successful heavier-than-air machine, the airplane, was invented in 1903 by two brothers, Orville and Wilbur Wright. They were not scientists, nor did they publish a scholarly paper in a scientific journal delineating their plans. Quite the contrary, the two brothers had a background in printing presses, bicycles, motors, and other machinery. Clearly, their background would not suggest they would invent the first airplane and lead humankind into space. However, their experience in machinery enabled them to build a small wind tunnel and collect the data necessary to sustain controlled air flight. From the beginning, the Wright brothers believed that the solution to controlled air flight lay hidden in pilot controls, rather than powerful engines. Based on their wind tunnel work, they invented what is now the standard method of all airplane controls, the three-axis control. They also invented efficient wing and propeller designs. It is likely that many in the scientific community in the beginning of the twentieth century would have considered aeronautics similar to the way the scientific community in the early part of the twenty-first century considers time travel—still something outside the fold of legitimate science. However, on December 17, 1903, at a small, remote airfield in Kitty Hawk, North Carolina, the two brothers made the first controlled, powered, and sustained heavier-than-air human flight. They invented the airplane. It was, of course, humankind's first step into the heavens.

I believe the invention of the airplane is a good analogy to where we are regarding time travel. We have some examples, namely, time dilation data, and a theoretical basis that suggests time travel is potentially real. However, we have not reached the "Kitty Hawk" moment. If Dr. Mallett makes his time machine work, and that is a big "if," numerous physicists will provide the theoretical foundation for its success, essentially erasing any errors that Dr. Mallett may have made in his calculations. He will walk as another great into the history of scientific achievement.

My point is a simple one. The line between scientific genius and scientific "crank" is a fine one. When Einstein initially introduced his special theory of relativity in 1905, he was either criticized or ignored. Few in the scientific community appreciated and understood Einstein's special theory of relativity in 1905. It took about fifteen years for the scientific community to begin to accept it. Einstein was aware of the atmosphere that surrounded him. In 1919, he stated in the *Times* of London, "By an application of the theory of relativity to the taste of readers, today in Germany I am called a German man of science, and in England I am represented as a Swiss Jew. If I come to be represented as a bête noire, the descriptions will be reversed, and I shall become a Swiss Jew for the Germans and a German man of science for the English!"

Dr. Mallett is on record predicting a breakthrough in backward time travel within a decade. Only time and experimental evidence will prove if his prediction becomes reality. Even if the Mallett time machine works, it would still represent only a baby step. We would still be a long way from human time travel, but we would be one step closer.

Conclusions from Chapter 5

- There is a theoretical basis in both Einstein's special and general theories of relativity that suggests forward and backward time travel is possible.

- Forward time travel (i.e., time dilation) has been demonstrated numerous times, especially using particle accelerators capable of accelerating a subatomic particle near the speed of light.

- No similar apparatus exists to demonstrate backward time travel (i.e., closed timelike curves).

- As of this writing, the only real time machine is a particle accelerator capable of accelerating a subatomic particle close to the speed of light, which results in time dilation (i.e., forward time travel).

- There is no consensus that the work of Dr. Mallett will result in a real time machine capable of sending subatomic particles backward in time via the creation of closed timelike curves. Therefore, we will need experimental evidence to prove its validity.

- The largest issues associated with time travel are:

  o Generating sufficient positive energy to accelerate sizable objects, for example a spacecraft, close to the speed of light, which would allow humans to experience noticeable time dilation. To date, the fastest manned spacecraft was *Apollo 10*, which obtained a speed of 25,000 miles per hour. We would need to go more than 13,000 times faster to travel in time (i.e., experience noticeable time dilation), which is about half the speed of light. The speed of light is approximately 186,000 miles per second.

o   Generating sufficient negative energy to enable traversable wormholes, or similar exotic time machines, to travel backward in time (i.e., closed timelike curves).

o   The engineering to build a real time machine, capable of human time travel, has proved incredibly difficult, or we would have already solved the engineering problems.

- The "Kitty Hawk" moment may be near, in the sense that true time machines capable of sending sizable objects, or even humans, may be developed during the twenty-first century:

o   We have taken a baby step in forward time travel via particle accelerators capable of accelerating a subatomic particle close to the speed of light, which results in time dilation.

o   We have not taken even a baby step in backward time travel, but respected physicists are dialoguing on its feasibility.

- The current situation regarding time travel technology is similar to where space travel technology was at the beginning of the twentieth century, and a breakthrough could come at any time, even from an unlikely source.

- Numerous physicists have expressed concerns regarding time travel paradoxes, which are thought experiments demonstrating causality (i.e., cause precedes effect) and reverse causality (i.e., effect precedes cause) violations. Other physicists have proposed solutions or thought experiments to address time travel paradoxes. It remains unclear whether time travel paradoxes present a serious obstacle to time travel.

The time machines proposed in this chapter have their roots in solid theory, including Einstein's special and general theories of relativity, and the existence equation conjecture. However, the gap between theory and engineering is incredibly large. We face numerous challenges. For example, we need to deal with the reverse causality issues and numerous other technical obstacles, which are the subject of our next chapter: "What Are the Obstacles to Time Travel?"

"If we could travel into the past, it's mind-boggling what would be possible. For one thing, history would become an experimental science, which it certainly isn't today. The possible insights into our own past and nature and origins would be dazzling. For another, we would be facing the deep paradoxes of interfering with the scheme of causality that has led to our own time and ourselves. I have no idea whether it's possible, but it's certainly worth exploring."

—**Carl Sagan**
***NOVA* interview, October 12, 1999**

CHAPTER 6

What Are the Obstacles to Time Travel?

There are enormous challenges associated with time travel. In this chapter, we will discuss the most significant engineering challenges, as well as the numerous philosophical issues, which are typically termed "time travel paradoxes." We will also discuss Stephen Hawking's chronology protection conjecture.

The engineering challenges to time travel

Let us start our discussion with the formidable engineering challenges to time travel.

- Time—There is no scientific consensus regarding a definition of time (see appendix 5 for more information). It is only natural to ask the question: How can we travel through something we do not scientifically understand? While it is true that Einstein's special theory of relativity provides insight into the nature of time, it is not a complete picture. For example, quantum mechanics does not provide an equivalent insight into the nature of time at the atomic and subatomic level. This compounds the problem of time travel. Will we have to wait for a theoretical breakthrough that explains time itself? Does the existence equation conjecture represent this breakthrough? It is hard to answer this last question until the existence equation conjecture is vetted via peer review.

- Energy—Without doubt, harnessing sufficient energy is one of the largest obstacles to time travel. For example, time dilation (i.e., forward time travel) is only noticeable when mass approaches a significant fraction of the speed of light or sits in a strong gravitational field. To date, we have been able to accelerate subatomic particles to a point where time dilation becomes noticeable. We have also been able to observe time dilation of a highly accurate atomic clock on a jet plane as it flies over the airport, which contains another atomic clock. Using sensitive instruments, we can measure time dilation. We have also been able to measure time dilation due to differences in the Earth's gravitational field. However, these differences are only evident using highly accurate atomic clocks. Our human senses are unable to detect a high mounted wall clock moving faster than our wristwatch, which gravitational time dilation predicts is occurring.

The fastest humankind has traveled is 25,000 miles per hour, using the *Apollo 10* spacecraft. The speed of light in a vacuum is approximately 186,000 miles per second. This means that a spacecraft would have to go about 13,000 times faster than *Apollo 10* for humans to experience noticeable time dilation, or a speed of about 90,000 miles per second, which is roughly half the speed of light. Today's science has not learned to harness the amount of energy required to accelerate a spacecraft to a velocity of 90,000 miles per second.

Let us consider a simple example to illustrate the amount of energy required to achieve the above velocity. If we have a mass of 1000 kilograms (i.e., 2204 pounds), and we want to accelerate it to 10% the speed of light, the resulting kinetic energy would be about 10^{17} (i.e., a 1 with 17 zeros after it) joules, whether you calculate the kinetic energy using Newton's classical formula or Einstein's relativistic formula for kinetic energy. To put this in perspective, it is more than twice the amount of energy of the largest nuclear bomb ever detonated. It would take a modern nuclear power plant about ten years to output this amount of energy.

The above example gives us a conceptual framework to understand the amount of energy that would be required to accelerate a sizable mass, 1000 kilograms, or 2204 pounds, to just 10% the speed of light. If we wish to accelerate the mass, for example, a spacecraft, to a greater percentage, the energy increases exponentially. For example, to accelerate to 20% the speed of light would require four times the amount of energy.

Today's engineering is unable to harness this level of energy. In the popular *Star Trek* television series and movies, the starship *Enterprise* is able to travel faster than the speed of light using a warp drive, by reacting matter with antimatter. Factually, there is almost no antimatter in the universe. This is one of the mysteries associated with the big bang science theory, which I discussed in my book, *Unraveling the Universe's Mysteries*. In theory, during the big bang, matter and antimatter should exist in equal quantities. Our observation of the universe, using our best telescopes, detects almost no antimatter. However, Fermi National Accelerator Laboratory (Fermilab) in Illinois is able to produce about fifty billion antiprotons per hour. This, though, is a miniscule amount compared to the amount needed to power a starship. According to Dr. Lawrence Krauss, a physicist and author of *The Physics of Star Trek*, it would take one hundred thousand Fermilabs to power a single lightbulb. In essence, we are a long way from using matter-antimatter as a fuel. In addition, the *Enterprise* was able to warp space. This provided a means to skirt around Einstein's well-established special theory of relativity, which asserts no mass can travel faster than the speed of light. There is no similar physical law that prohibits space from expanding faster than the speed of light. If we are able to manipulate space, similar to our discussion of the Alcubierre drive in the previous chapter, then scientifically the spacecraft could collapse space in front of it and expand space behind it. However, the Alcubierre drive requires negative energy. Today's science is unable to create and harness negative energy in any significant way.

Therefore, topping our list of major scientific obstacles regarding time travel is generating huge amounts of energy, in either positive or negative form.

- Building a traversable wormhole—Prominent physicist Stephen Hawking believes that time travel using a traversable wormhole may one day be possible. Unfortunately, most physicists believe that a traversable wormhole will require negative energy, which we discussed in the previous chapter. Today's engineering is unable to generate any significant amount of negative energy. However, as we discussed in the previous chapter, some solutions to Einstein's equations of general relativity suggest it may be possible to build a traversable wormhole using huge amounts of conventional positive energy. Dr. Mallett, for example, is attempting to twist spacetime using a laser (i.e., conventional positive energy), as we discussed in the previous chapter. However, most of the scientific community remains dubious that Dr. Mallett's approach will result in actually twisting spacetime due to the energy required.

To build a traversable wormhole, we face significant challenges. Today's engineering does not know how to generate significant amounts of negative energy, or even significant amounts of positive energy. Building a wormhole type of time machine, similar to the approach of Dr. Mallett, appears beyond today's engineering capabilities.

- Engineering a time machine—The theoretical science of forward time travel (i.e., time dilation) has been known for over a century. The theoretical science of backward time travel (i.e., closed timelike curves) has been known for over sixty years. However, building a time machine capable of performing time dilation, or closed timelike curves, has proved a formidable engineering task. To date, the best engineers and scientists have only been able to build particle accelerators, like the Large Hadron Collider completed in 2008, to accelerate subatomic particles to a high fraction of the speed of light and observe time dilation. No similar machine exists to produce closed timelike curves. Dr. Mallett's spacetime twisting by light (STL) machine is as close as we have come. To date, the Mallett time machine has not demonstrated closed timelike curves, and several well-respected physicists doubt its ability to do so. This suggests the engineering task to build a time machine capable of sending humans, or even information, forward or backward in time is an extremely difficult task. The task is so difficult that, to date, it has eluded the best engineering minds.

The above obstacles are formidable, but then again, so was space travel at the beginning of the twentieth century. The challenges to time travel do not represent insurmountable obstacles. We are not violating any known physical law when we consider constructing a wormhole, generating negative energy, or harnessing huge amounts of positive energy. In the fullness of time, perhaps measured in centuries, science will address and overcome the challenges. As history has illustrated, today's science fiction becomes tomorrow's science fact.

The next challenges to time travel are real mind benders. They force us to come to terms with the essence of causality (i.e., cause precedes effect) and reverse causality (i.e., effect precedes the cause). They are time travel paradoxes.

Time travel paradoxes

What is a time travel paradox? It is an occurrence that apparently violates some aspect of causality typically associated with time travel.

Although there are numerous time travel paradoxes, let us explore some of the most famous ones:

- The grandfather paradox—Science fiction writer René Barjavel, in his 1943 book, *Le Voyageur Imprudent* (*Future Times Three*), originally proposed the grandfather paradox. It goes something like this. A person goes back in time and meets his grandfather before his grandfather meets his grandmother. The person in some way interferes with his grandfather meeting his grandmother. Consequently, the grandfather and grandmother never meet. The question becomes, what happens to the person? In theory, the person will never be born.

  Is this just some illogical premise, similar to asserting that a square circle exists? Most of the scientific community considers it a valid concern regarding causality violations due to time travel. Some physicists believe that it actually presents a barrier to time travel. However, numerous theories exist to resolve time travel paradoxes. We will discuss those theories in the next section, but first let us explore another famous paradox.

- The twin paradox—The is one of the most famous time travel paradoxes. It goes something like this: On Earth live a pair of twins. They are almost the same age, differing only by the order in which they were born. One twin boards a spacecraft capable of traveling near the speed of light. In the spacecraft, the twin embarks on a one-year journey, measured by the clock within the spacecraft. During the one-year journey, the spacecraft travels at 99.94% the speed of light. When the spacecraft returns to Earth, the twin on the spacecraft has aged one year, but learns his twin has aged almost thirty years. Although the example is fictitious, the science is real. The twin paradox has been experimentally verified using highly accurate atomic clocks, one on a jet plane and the other at the airport. There have been many variations of the twin paradox. The scientific community considers it a valid effect of Einstein's special theory of relativity regarding time dilation.

Our next paradox has to do with the future changing the present or the past. The effect has been known for well over a hundred years. It continues to this day to be a topic of discussion.

- The double-slit experiment—There are numerous versions of the double-slit experiment. In its classic version, a coherent light source, for example a laser, illuminates a thin plate containing two open parallel slits. The light passing through the slits causes a series of light and dark bands on a screen behind the thin plate. The brightest bands are at the center, and the bands become dimmer the farther they are from the center. See figure 3 below:

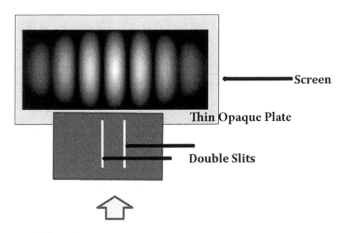

Figure 3: Monochromatic light passes through the double slits and creates an interference pattern on the screen behind the slits.

The series of light and dark bands on the screen would not occur if light were only a particle. If light consisted of only particles, we would expect to see only two slits of light on the screen, and the two slits of light would replicate the slits in the thin plate. Instead, we see a series of light and dark patterns, with the brightest band of light in the center, and tapering to the dimmest bands of light at either side of the center. This is an interference pattern and suggests that light exhibits the properties of a wave. We know from other experiments—for example, the photoelectric effect (see glossary), which I discussed in my first book, *Unraveling the Universe's Mysteries*—that light also exhibits the properties of a particle. Thus, light exhibits both particle- and wavelike properties. This is termed the dual nature of light. This portion of the double-slit experiment simply exhibits the wave nature of light. Perhaps a number of readers have seen this experiment firsthand in a high school science class.

The above double-slit experiment demonstrates only one element of the paradoxical nature of light, the wave properties. The next part of the double-slit experiment continues to puzzle scientists. There are five aspects to the next part.

1. Both individual photons of light and individual atoms have been projected at the slits one at a time. This means that one photon or one atom is projected, like a bullet from a gun, toward the slits. Surely, our judgment would suggest that we would only get two slits of light or atoms at the screen behind the slits. However, we still get an interference pattern, a series of light and dark lines, similar to the interference pattern described above. Two inferences are possible:

    a. The individual photon light acted as a wave and went through both slits, interfering with itself to cause an interference pattern.

     b. Atoms also exhibit a wave-particle duality, similar to light, and act similarly to the behavior of an individual photon light described (in part a) above.

2. Scientists have placed detectors in close proximity to the screen to observe what is happening, and they find something even stranger occurs. The interference pattern disappears, and only two slits of light or atoms appear on the screen. What causes this? Quantum physicists argue that as soon as we attempt to observe the wavefunction of the photon or atom, it collapses. Please note, in quantum mechanics, the wavefunction describes the propagation of the wave associated with any particle or group of particles. When the wavefunction collapses, the photon acts only as a particle.

3. If the detector (in number 2 immediately above) stays in place but is turned off (i.e., no observation or recording of data occurs), the interference pattern returns and is observed on the screen. We have no way of explaining how the photons or atoms know the detector is off, but somehow they know. This is part of the puzzling aspect of the double-slit experiment. This also appears to support the arguments of quantum physicists, namely, that observing the wavefunction will cause it to collapse.

4. The quantum eraser experiment—Quantum physicists argue the double-slit experiment demonstrates another unusual property of quantum mechanics, namely, an effect termed the quantum eraser experiment. Essentially, it has two parts:

     a. Detectors record the path of a photon regarding which slit it goes through. As described above, the act of measuring "which path" destroys the interference pattern.

b. If the "which path" information is erased, the interference pattern returns. It does not matter in which
order the "which path" information is erased. It can
be erased before or after the detection of the photons.

This appears to support the wavefunction collapse theory,
namely, observing the photon causes its wavefunction to collapse and assume a single value.

5. If the detector replaces the screen and only views the atoms or
photons after they have passed through the slits, once again,
the interference pattern vanishes and we get only two slits of
light or atoms. How can we explain this? In 1978, American
theoretical physicist John Wheeler (1911–2008) proposed that
observing the photon or atom after it passes through the slit
would ultimately determine if the photon or atom acts like a
wave or particle. If you attempt to observe the photon or atom,
or in any way collect data regarding either one's behavior, the
interference pattern vanishes, and you only get two slits of
photons or atoms. In 1984, Carroll Alley, Oleg Jakubowicz, and
William Wickes proved this experimentally at the University of
Maryland. This is the "delayed-choice experiment." Somehow,
in measuring the future state of the photon, the results were
able to influence their behavior at the slits. In effect, we are
twisting the arrow of time, causing the future to influence the
past. Numerous additional experiments confirm this result.

Let us pause here and be perfectly clear. Measuring the
future state of the photon after it has gone through the
slits causes the interference pattern to vanish. Somehow,
a measurement in the future is able to reach back into
the past and cause the photons to behave differently.

In this case, the measurement of the photon causes its wave nature to vanish (i.e., collapse) even after it has gone through the slit. The photon now acts like a particle, not a wave. This paradox is clear evidence that a future action can reach back and change the past.

To date, no quantum mechanical or other explanation has gained widespread acceptance in the scientific community. We are dealing with a time travel paradox that illustrates reverse causality (i.e., effect precedes cause), where the effect of measuring a photon affects its past behavior. This simple high-school-level experiment continues to baffle modern science. Although quantum physicists explain it as wavefunction collapse, the explanation tends not to satisfy many in the scientific community. Irrefutably, the delayed-choice experiments suggest the arrow of time is reversible and the future can influence the past.

- The grandchild paradox—I am introducing a new term regarding time travel paradoxes, namely, the "grandchild paradox." This term is intended to mirror the grandfather paradox. In any mirror, the image is reversed. The grandchild paradox is the reverse of the grandfather paradox. The grandchild paradox refers to any situation involving reverse causality. Any situation, real or imagined, that reverses the arrow of time and allows the future to influence the past, may be considered a grandchild paradox. Therefore, the double-slit delayed-choice experiment described above is an example of the grandchild paradox.

- The Fermi paradox—This paradox was initially suggested by Italian theoretical and experimental physicist Enrico Fermi

(1901–1954). The paradox was originally suggested in the context of extraterrestrial spacecraft (i.e., UFOs). The story goes something like this. In 1950, Enrico Fermi was employed by Los Alamos National Laboratory. During a casual conversation at lunch with colleagues, the subject of UFOs came up. There had been many recent UFO sightings. However, there was scant real evidence reported. As the story is told, Fermi suddenly exclaimed, "Where are they?" Fermi then proceeded to make a series of rapid calculations. Fermi was well known for his ability to make calculations using first principle and minimal data. According to Fermi's calculation, the Earth should have been visited by UFOs, historically, many times. The Fermi paradox, then, has to do with the apparent contradiction between high estimates of the probability that an event is possible, and the apparent lack of evidence.

In this case, the Fermi paradox would address the question, "If time travel is possible, where are all the time travelers?" This question is addressed by simply understanding the "traversable wormhole" or the "Mallett time machine." From these examples, we can assert it is not possible, based on fundamental principles, to go back to a time before the time machine is invented. Since we have no verifiable scientific evidence that time travelers have visited us at any time, we can assert a time machine capable of allowing humans to time travel has not yet been invented.

There is a laundry list of time travel paradoxes. To discuss them all would be a book in itself. The paradoxes above are sufficient to illustrate causality and reverse causality issues. It is important to note that the time travel paradoxes are not simply in the category of thought experiments. Numerous time travel paradoxes, like the twin paradox and the double-slit delayed-choice paradox, are experimental facts. They are real.

Let's now consider another obstacle to time travel. It has been proposed by the most famous scientist alive today, Stephen Hawking.

Chronology protection conjecture

In 1992, Stephen Hawking published a paper called "Chronology Protection Conjecture." In it, Hawking argues the laws of physics do not allow the appearance of closed timelike curves.

Dr. Hawking bases his arguments on semiclassical gravity, concluding that vacuum fluctuations would drive the energy density at the boundary (i.e., the point where closed timelike curves occur) of the time machine to infinity. This, of course, would destroy the time machine, making it impossible to use.

Dr. Hawking published his chronology protection conjecture much like a law of physics, which created something of a backlash. Many physicists questioned: Is this constraint on closed timelike curves truly a physical law? The argument opponents raised centered on Dr. Hawking's use of semiclassical gravity. The opponents state that the chronology protection conjecture would require a full theory of quantum gravity (i.e., a theoretical field of physics that is attempting to unify quantum mechanics with Einstein's general relativity) to prove its validity. Today, there is no widely accepted theory of quantum gravity.

The debate on the validity of the chronology protection conjecture is heated. In fact, Li-Xin Li, a Princeton physicist quoted in Dr. Kaku's book, *Parallel Worlds*, published an "anti-chronology protection conjecture," stating, "There is no law of physics preventing the appearance of closed timelike curves."

In 1998, this significant backlash caused Dr. Hawking to acknowledge that portions of the criticism are valid.

Conclusions from Chapter 6

There are formidable challenges to building a time machine. The challenges fall into three broad groups:

1. Engineering challenges—There is no widely accepted scientific definition of time. Obviously, this raises a valid question: How can we travel through something we do not scientifically understand? In addition, today's engineering is unable to create sufficient positive or negative energy, thought by both engineers and physicists, required to twist spacetime and create traversable wormholes or spacecraft capable of warping space and/or traveling near the speed of light.

2. Time travel paradoxes—Numerous time travel paradoxes suggest causality (i.e., cause precedes effect) and reverse causality (i.e., effect precedes cause) issues caused by time travel. Several time travel paradoxes are experimental facts, as discussed above. They are real. We will discuss if they present a significant barrier to time travel in the next chapter.

3. Theoretical challenges—The most formidable theoretical challenge to human time travel is Stephen Hawking's chronology protection conjecture. If Dr. Hawking is correct in his conjecture, it rules out human time travel. However, Dr. Hawking has received significant criticism on his conjecture and has acknowledged that some of the criticism is valid. Because Dr. Hawking was unable to formulate his theory using quantum gravity, many in the scientific community do not think the chronology protection conjecture rests on a solid scientific foundation.

At this point, it is valid to ask this question: Will we overcome the obstacles to time travel? Obviously, they are formidable. However, the obstacles to put a man on the moon were also formidable, and humankind overcame them. Is it reasonable to believe, in a similar fashion, we will overcome the obstacles associated with time travel? The scientific community has some specific views that address this question. Their views may surprise you.

"Obstacles cannot crush me; every obstacle yields to stern resolve."

—**Leonardo da Vinci (1452 – 1519)**

CHAPTER 7

Will We Overcome the Obstacles to Time Travel?

We will start our discussion by examining methods to overcome the scientific aspects of time travel.

Addressing the scientific challenges

- Time—Achieving a scientific definition of time is obviously the first major obstacle. The two most successful scientific theories, Einstein's theories of relativity and the theory of quantum mechanics, do not combine to give us a unified scientific definition of time. This raises a reasonable question: How can we move through something we cannot scientifically define?

- Energy—Harnessing energy is a multipronged challenge. There are numerous forms of energy, which measure the ability of one object or system to do work on another object or system. Here are the basic forms of energy:

  o Kinetic energy—This is the energy associated with an object's movement in spacetime.

  o Potential energy—This is the energy associated with a body due to its position relative to others, stresses within itself, electric charge, and potentially other known and unknown factors.

  o Thermal, or heat energy—This is the energy associated with the collective, microscopic, kinetic, and potential energy of the atoms and molecules of a body.

  o Chemical energy—This is any form of energy resulting from a chemical reaction.

  o Electrical energy—This is the energy associated with the flow of electrons.

  o Electrochemical energy—This refers to any chemical reaction resulting in the flow of electrons.

  o Electromagnetic energy—This refers to the energy associated with a photon (i.e., a discrete packet) of light.

o Solar energy—I have added this because many people may not make the connection that this is electromagnetic energy, which is discussed above.

o Sound energy—This refers to the kinetic and potential energy associated with compressed air molecules.

o Nuclear energy—This refers to any type of energy resulting from a nuclear reaction.

Most of us are familiar with the above types of energy. For example, we know that if we throw a baseball at a window, it will likely break the window due to the baseball's kinetic energy. We know that our house lamps use electricity (i.e., flow of electrons) to light a room. My point is that most of us are familiar with and even use many of the above types of energy in our everyday lives. The challenge is creating and harnessing large amounts of an energy source. For example, we have not been able to create and harness enough energy to accelerate a spacecraft close to the speed of light, which would be required for noticeable time dilation (i.e., forward time travel).

What level of energy is it going to take to accelerate a spacecraft near the speed of light, or warp spacetime (i.e., travel faster than light)? The real answer is no one knows. However, we know it is more than we have been able to create and harness to date. If we borrow our answer from science fiction, it is going to require we learn to make matter-antimatter reactors, or harness the complete energy of a star, like our sun. Perhaps we will need to learn to create large quantities of negative energy. The list regarding the type of energy we need, the amount we need, and the method to generate it is almost unimaginable. We do not know how much or even what kind of energy we need to warp spacetime to enable time travel, and even if we did, we do not know how to generate it.

If you think about it, humankind has been able to create and harness energy in small quantities dating back to the time the first human learned to make fire. However, what we consider large quantities of energy, such as the amount of energy we use to launch the space shuttle, is still a relatively small amount of energy on a cosmic basis. For example, according to NASA, if we just wanted to visit our nearest neighbor star, at sublight speeds, it would take a space shuttle type craft burning fuel at full power about fifty years to reach it. That is a lot of fuel, and it is not practical. We do not know how to do it.

Based on today's science, no technology has surfaced offering the potential to generate sufficiently large quantities of energy to effect near-the-speed-of-light space travel, where time dilation would be noticeable, let alone generate the amount of energy required for significant time travel. This is not from lack of trying. NASA has compiled a list of emerging technologies on their official website (http://www.nasa.gov/centers/glenn/technology/warp/possible.html) that they feel may offer the potential for a propulsion breakthrough. Their list is below:

- 2001 BPP (Breakthrough Propulsion Physics)-Sponsored Papers presented at the BPP Sessions of the July 2001 Joint Propulsion Conference in Salt Lake City, Utah (intended for technical audiences).

- 1996 Eberlein: Theory suggesting that the laboratory observed effect of sonoluminescence is an extraction of virtual photons from the electromagnetic zero point fluctuations.

- 1994 Alcubierre: Theory for a faster-than-light "warp drive" consistent with general relativity.

- 1994 Haisch, Rueda, and Puthoff: Theory suggesting that inertia is a consequential effect of the vacuum electromagnetic zero point fluctuations.

- 1992 Podkletnov and Nieminen: Report of superconductor experiments with anomalous results—evidence of a possible gravity shielding effect.

- 1989 Puthoff: Theory extending Sakharov's 1968 work to suggest that gravity is a consequential effect of the vacuum electromagnetic zero point fluctuations.

- 1988 Herbert: Book outlining the loopholes in physics that suggest that faster-than-light travel may be possible.

- 1988 Morris and Thorne: Theory and assessments for using wormholes for faster-than-light space travel.

It is an impressive list, and it suggests that NASA is actively engaged in the study of breakthrough propulsion physics (BPP). However, according to NASA, they are a long way from knowing whether any of these developments can lead to the desired propulsion breakthroughs.

While I am not a conspiracy theory advocate, I have held a secret clearance. I have worked on programs to build some of the most advanced spacecraft ever developed. Obviously, NASA did not publish this work on their website. Therefore, I wonder, what are they pursuing but not telling us?

However, NASA does make its goal known, stating on their website, "These emerging ideas are all related in some way to the physics goals for practical interstellar travel; controlling gravitational or inertial forces, traveling faster-than-light, and taking advantage of the energy in the space vacuum. Even though the physics has not yet matured to where 'space drives' or 'warp drives' can be engineered, individuals throughout the aerospace community and across the globe have been tracking these and other emerging clues." Consistent with their goal, NASA lists peer-reviewed scientific papers resulting from NASA's Breakthrough Propulsion Physics (BPP) project sponsorship. Here is that list from NASA BPP website, which they provide in reverse numerical order enabling the most recent publication to appear at the top: (http://www.grc.nasa.gov/WWW/bpp/JournalList.html):

16. Zampino, Edward J. 2006. "Warp-Drive Metrics and the Yilmaz Theory." *Journal of the British Interplanetary Society*, 59: 226–229.

15. Millis, M. G. 2005. "Assessing Potential Propulsion Breakthroughs." *Annals of the New York Academy of Sciences*, 1065: 441–461 (December).

14. Deck, Robert, Jacques Amar, and Gustave Fralick. 2004. "Nuclear size correction to the energy levels of single-electron and -muon atoms." *Journal of Physics B: Atomic, Molecular and Optical Physics*, 38: 2173–2186.

13. Maclay, J., and R. Forward. 2004. "A Gedanken spacecraft that operates using the quantum vacuum (adiabatic Casimir effect)." *Foundations of Physics*, 34: 477–500.

12. Milonni, P. W., and J. Maclay. 2003. "Quantized-Field Description of Light in Negative-Index Media." *Optics Communications*, 228: 161–165.

11. Villarreal, C.; Esquivel-Sirvent, R.; Cocoletzi, G. H. 2002. "Modification of Casimir Forces due to Band Gaps in Periodic Structures." *International Journal of Modern Physics A*, 17: 798–803.

10. Mochan, W. L., Esquivel-Sirvent, and Villarreal. 2002. "On Casimir Forces in Media with Arbitrary Dielectric Properties." *Revista Mexicana de Fisica*, 48: 339.

9. Esquivel-Sirvent, R., Villarreal, Mochan, and Cocoletzi. 2002. "Casimir Forces in Nanostructures." *Physica Status Solidi (b)*, 230: 409–413.

8. Maclay, G. J., Fearn, and Milonni. 2001. "Of some theoretical significance: implications of Casimir effects." *European Journal of Physics*, 22: 463–469.

7. Esquivel-Sirvent, R., Villarreal, and Cocoletzi. 2001. "Superlattice-mediated tuning of Casimir forces." *Physical Review A*, 64: 052108-1 to 052108-4.

6. Segev, B., Milonni, Babb, and Chiao. 2000. "Quantum noise and superluminal propagation." *Physical Review A*, 62: 0022114-1 to 0022114-15.

5. Maclay, G. J. 2000. "Analysis of zero-point electromagnetic energy and Casimir forces in conducting rectangular cavities." *Physical Review A*, 61: 052110-1 to 052110-18.

4. Mojahedi, M., Schamiloglu, Kamil, and Malloy. 2000. "Frequency Domain Detection of Superluminal Group

Velocities in a Distributed Bragg Reflector." *IEEE Journal of Quantum Electronics*, 36: 418–424.

3. Mojahedi, M., Schamiloglu, Hegeler, and Malloy. 2000. "Time-domain detection of superluminal group velocity for single microwave pulses." *Physical Review E*, 62: 5758–5766.

2. Millis, M. G. 1999. "NASA Breakthrough Propulsion Physics Program." *Acta Astronautica*, 44: 175–82 (Eqv. NASA TM-1998-208400).

1. Millis, M. G. 1997. "Challenge to Create the Space Drive." *AIAA Journal of Propulsion and Power*, 13: 577–582.

Once again, I want to stress that any work classified as "secret" or higher is not going to be on the list. There is, obviously, much more scientific data than what is publicly available to the scientific community.

Can we infer what data NASA might have, but is withholding? I think we can get close by asking one question: What is it we need but do not have to create an interstellar spacecraft propulsion engine? Here is the short list, based on my research:

1. A matter-antimatter propulsion system—No reaction is more efficient in converting mass to energy than the interaction of matter with antimatter. Based on results published in 2010 by CERN physicists, matter-antimatter reactions are 99% efficient in creating electromagnetic energy. The 1% left over after the reaction goes to completion is matter. In contrast, in our first nuclear weapon, Little Boy, also known as the Hiroshima bomb, only 1% of the 112 pounds of U-235 uranium underwent fission,

releasing the energy associated with the mushroom cloud that leveled Hiroshima. The remainder was scattered by the blast, but did not add to the blast's energy. Therefore, Little Boy was only 1% efficient. The efficiency of today's nuclear weapons is higher, but the efficiency levels are classified. Imagine, though, the amount of destruction that would have resulted if Little Boy were, for example, 10% efficient. This will give some idea of the propulsion capabilities of a 99% efficient matter-antimatter engine.

Is it possible that NASA has a matter-antimatter engine? My judgment is they probably do not have one, but they are working on it. The reason I judge they do not have one is that antimatter is extremely difficult to make. The best way to produce antimatter is using particle accelerators, like Fermi National Accelerator Laboratory (FNAL) and the European Laboratory for Particle Physics (CERN). The published number for the world's production of antimatter in 2008 was in the order of several picograms (1 picogram = 1/1,000,000,000,000 gram). The production rate is slowly increasing, and the current production rate is in the order of 10 nanograms (1 nanogram = 1/1,000,000,000 gram) per year, which is about a thousand times more than 2008. However, it still represents a minuscule amount of antimatter. Therefore, the first breakthrough would need to be the ability to make and store significant quantities of antimatter. In a side note, the United States Air Force has been funding antimatter research since the Cold War (about the 1960s), based on its potential as a weapon of mass destruction and its ability to act as a rocket fuel. Unfortunately, humankind has proven particularly adept at making advances in weapons. I have no reason to believe that antimatter would be an exception.

2. A method to create a significant quantity of negative energy—
Some physicists argue that light travels faster in a negative
energy vacuum that our normal positive energy vacuum. They
propose that a spacecraft could travel faster than the speed of
light if it could surround itself with a negative energy vacuum.
In "Borrowed Time: Interview with Michio Kaku" in *Scientific American* (November 23, 2003), prominent physicist and
author Michio Kaku stated, "It would take a civilization far
more advanced than ours, unbelievably advanced, to begin to
manipulate negative energy to create gateways to the past. But
if you could obtain large quantities of negative energy—and
that's a big 'IF'—then you could create a time machine that
apparently obeys Einstein's equation and perhaps the laws of
quantum theory."

Many physicists believe the Casimir effect, discussed in chapter 4, is proof that negative energy naturally occurs. Physicists
are exploiting the Casimir effect to create negative energy. Have
they succeeded? The answer is likely "top secret."

3. A method to find or create negative mass—Although we do
not know of any naturally occurring negative mass, Hermann
Bondi theorized its existence in 1957, as discussed in chapter
4. Even NASA has publicly stated on their website that a negative mass drive is theoretically possible: "It has been shown that
it is theoretically possible to create a continuously propulsive
effect by the juxtaposition of negative and positive mass and that
such a scheme does not violate conservation of momentum or
energy." However, they make no statement that they have found
or created negative mass. To date, NASA only states, "Methods
have been suggested to search for evidence of negative mass in
the context of searching for astronomical evidence of wormholes." To my mind, it makes sense that NASA and others in

the scientific community either have or are attempting to find or create negative mass. I believe this because once something is theorized to exist, the scientific community is naturally drawn to look for it. Has NASA, or any defense contractor, found or created negative mass? Once again, the answer would be "top secret."

The above is my short list of technologies required for an interstellar spacecraft propulsion engine. Have there been significant breakthroughs? Do we have an interstellar spacecraft engine? Someone knows, but the answers to those questions are likely above "top secret." The levels above top secret are either "sensitive compartmented information" (SCI) or "special access program" (SAP), and the "need to know" rule is rigorously enforced. That means that even if your classification is sensitive compartmented information, you only get access to the information necessary for you to do your job. My point is that the information regarding the science and engineering of an interstellar propulsion engine is not likely to find its way into a scientific journal due to its national security sensitivity.

Resolving time travel paradoxes

Do time travel paradoxes spell doom to time travel? The short answer is no. Many in the scientific community do not think time travel paradoxes present an insurmountable barrier to time travel. Many physicists have suggested solutions to time travel paradoxes. In fact, discussing them all would result in a book. I will discuss the major ones. For the sake of convenience, I have divided them into four categories:

1. Multiverse hypothesis—The multiverse hypothesis argues that time travel paradoxes are real, but they lead to alternate

realities. The most famous theory in this category is American physicist Hugh Everett's many-worlds interpretation (MWI) of quantum mechanics. According to Everett (1930–1982), certain observations in reality are not predictable absolutely by quantum mechanics. Instead, there is a range of possible observations associated with physical phenomena, and each is associated with a different probability. Everett's interpretation is that each possible observation corresponds to a different universe, hence the name "many-worlds."

Let us consider a simple example. If you toss a coin in the air, it can come down heads or tails. The probability of getting heads is equal to the probability of getting tails. If you toss the coin, and it comes down heads, then there is another you, in another universe, who observes tails. This sounds like science fiction. However, according to a poll published in *The Physics of Immortality* (1994), 58% of scientists believe the many-world interpretation of quantum mechanics is true, 13% are on the fence (undecided), 11% have no opinion, and 18% do not believe it. Among the believers are Nobel laureates Murray Gell-Mann and Richard Feynman, and world-famous physicist/cosmologist Stephen Hawking.

In our everyday reality, many of us would reject the many-world interpretation of quantum mechanics because we do not experience it directly. However, let me point out, we do not experience the individual atoms of a book when we hold it. Yet, we know from sophisticated experimental analysis that the book is a collection of atoms. Unfortunately, in the strange world of quantum mechanics, our intuition and experience rarely serve us. I leave it to you to formulate your own conclusions.

2. Timeline-protection hypothesis—The timeline-protection hypothesis asserts that it is impossible to create a time travel paradox. For example, if you travel back in time and attempt to prevent your grandfather from meeting your grandmother, you fail every time. If you attempt to shoot yourself through a wormhole, the gun jams, or something else happens, which prevents you from changing the past. Several other paradox resolutions fit under this category. They are:

   a. The Novikov self-consistency principle, suggested by Russian physicist Igor Dmitriyevich Novikov in the mid-1980s, which asserts anything a time traveler does remains consistent with history. For example, if you travel to the past and attempt to keep your grandfather from meeting your grandmother, something interferes with any attempt you make, causing you to fail in the attempt. In other words, the time traveler is unable to change history.

   b. The self-healing hypothesis theory, which states that whatever a time traveler does to alter the present by traveling to the past sets off another set of events to cause the present to remain unchanged. For example, if you attempt to prevent Abraham Lincoln's assassination, you may succeed in preventing John Wilkes Booth from carrying out the assassination only to find someone else assassinated Lincoln. In essence, time heals itself.

3. Timeline-corruption hypothesis—The timeline-corruption hypothesis suggests that time paradoxes are inevitable and unavoidable. Any time travel to the past creates minute effects that inevitably alter the timeline and cause the future to change.

For example, if you inadvertently step on an ant in the past, it changes the future. Popular science fiction literature calls this the "butterfly effect," namely, that the flutter of a butterfly's wings in Africa can cause a hurricane in North America. Under this theory, anything you do will have a consequence. It may be small and benign. Alternatively, it may be large and disastrous. The destruction-resolution hypothesis fits in this category. It holds that anything a time traveler does resulting in a paradox destroys the timeline, and even the universe. Obviously, if the destruction-resolution hypothesis is true, any time travel would be disastrous. However, I doubt the validity of the destruction-resolution hypothesis, since we are able to perform time dilation (i.e., forward time travel) experiments with subatomic particles using particle accelerators.

4. Choice timeline hypothesis—The choice timeline hypothesis holds that if you choose to travel in time, it is predestined, and history instantly changes. This implies you can time travel to the future and leave an item there that you will need sometime in the future. It will be there for you when the future becomes the present. For example, assume you are in New York City, and someone is about to assault you. You have no escape or means of protection. According to the choice timeline hypothesis, you can use your time machine to travel to the future. You hide a gun near the place where the assault is about to occur. When the assault occurs, you retrieve the hidden gun and scare off the attacker.

There are numerous other time-paradox resolution hypotheses. Most fall under one of the above categories, or are not as popular as the above. I left them out in the interest of clarity and brevity. The four categories above give us a reasonable framework to understand the major time-paradox resolution theories, and the current thinking regarding their impact on the timeline.

The majority of the scientific community does not think time paradoxes inhibit time travel. For example, Kip Thorne, an American theoretical physicist and professor of theoretical physics at the California Institute of Technology until 2009, argues that time paradoxes are imprecise thought experiments which can be resolved by numerous consistent solutions. The scientific consensus appears to be that time paradoxes may or may not occur, but they do not exclude the possibility of time travel. This position appears validated by the time dilation (i.e., forward time travel) experiments routinely performed using particle accelerators.

Evaluating the chronology protection conjecture

Most of the scientific community agrees that time travel is theoretically possible, based on Einstein's special and general theories of relativity. However, world-famous cosmologist and physicist Stephen Hawking published a 1992 paper, "Chronology Protection Conjecture," in which he stated the laws of physics do not allow the appearance of closed timelike curves. Since its publication, the chronology protection conjecture has been significantly criticized. Most of the criticism centered on Dr. Hawking's use of semiclassical gravity, versus using quantum gravity, to make his arguments. Dr. Hawking acknowledged, in 1998, that portions of the criticism are valid.

However, not to take sides on this issue, I feel compelled to point out that the two fundamental pillars of modern science, namely, general relativity and quantum mechanics, are incompatible. This placed Dr. Hawking in a difficult position regarding the use of gravity in writing the chronology protection conjecture. General relativity and quantum mechanics do not come together to provide a quantum gravity theory. This argues that we still do not have the whole picture, which makes it difficult to completely rule out Dr. Hawking's chronology protection conjecture.

Currently, there is no widespread consensus on any theory that unifies general relativity with quantum mechanics. If such a theory existed, it would be the theory of everything (TOE) and would provide us with a quantum gravity theory. Highly regarded physicists, such as Stephen Hawking, believe M-theory (i.e., membrane theory), which is the most comprehensive string theory, is a candidate for the theory of everything. However, there is significant disagreement in the scientific community. Many physicists argue that M-theory is not experimentally verifiable, and on that basis is not a valid theory of science. However, to be fair to all sides, Einstein's special theory of relativity, published in 1905, was also not experimentally verifiable for years. Today, most of the scientific community views the special theory of relativity as science fact, having withstood over one hundred years of scientific investigation. The scientific community, which didn't really know what to make of the special theory of relativity in 1905, hails it now as the "gold standard" of theories, arguing that other theories must measure up to the same standards of rigorous investigation. I think science is better served by a more moderate position. In this regard, I agree with prominent physicist and author Michio Kaku, who stated in Nina L. Diamond's *Voices of Truth* (2000), "The strength and weakness of physicists is that we believe in what we can measure. And if we can't measure it, then we say it probably doesn't exist. And that closes us off to an enormous amount of phenomena that we may not be able to measure because they only happened once. The Big Bang is an example. That's one reason why they scoffed at higher dimensions for so many years. Now we realize that there's no alternative."

In essence, we need to keep an open mind, regardless of how bizarre a scientific theory may first appear. However, we need to balance our open-mindedness with experimental verification. This, to my mind, is how science advances.

Building a real time machine

Scientifically speaking, we already have built a real time machine. A modern particle accelerator, such as the CERN Large Hadron Collider, is a real time machine. It can accelerate subatomic particles close to the speed of light and cause time dilation (i.e., forward time travel). Unfortunately, we do not have a similar machine that is able to cause subatomic particles to experience closed timelike curves (i.e., backward time travel). However, Dr. Mallett's spacetime twisting by light (STL) machine, discussed in chapter 5, may eventually result in such a backward time travel machine. This, though, is highly speculative. As mentioned previously, there is no scientific consensus that Dr. Mallett's device will work.

Although we do not have a real time machine capable of enabling human time travel, I have attempted to capture, via research, what the scientific community believes is necessary to develop real time travel for humans. There are two major components:

1. Forward time travel requires a machine, like a spacecraft, capable of accelerating to velocities extremely close to the speed of light. This will result in time dilation (i.e., forward time travel). The critical challenge is developing an energy source to power the machine. Based on all publicly available information, most of the interest and money to develop the energy source is targeting the generation and storage of antimatter. As previously mentioned, matter-antimatter reactions are 99% efficient. (Note: It is not 100% efficient, based on data provided by scientists at CERN in 2010.) In fact, one gram of antimatter reacting with one gram of matter can generate as much energy as twenty-three space shuttle external fuel tanks.

Alternately, if science is able to create negative mass, NASA has shown that it is theoretically possible to juxtapose negative mass with positive mass and create a propulsion source. Unfortunately, finding or creating negative mass has proved problematic. The standard model of particle physics does not predict a negative mass particle. When the standard model does not predict a particle, it is likely that it does not exist. On the other hand, numerous physicists argue that the region between the plates of a Casimir effect represents negative energy. This would imply negative mass if Einstein's famous mass-energy equivalence equation (i.e., $E = mc^2$) continues to be valid for negative mass/energy. Unfortunately, this leaves the question of negative mass's existence unresolved.

2. Creating closed timelike curves (i.e., backward time travel) requires manipulating the fabric of spacetime. For example, a traversable wormhole would require that one end of the wormhole be accelerated close to the speed of light, such that spacetime is twisted resulting in a closed timelike cure. Numerous well-respected physicists, such as Stephen Hawking and Kip Thorne, theorize this is going to require negative energy or negative mass (i.e., an exotic energy source). As discussed above, the whole area of creating negative energy or negative mass is problematic.

Until we develop a fuel, such as antimatter, or an exotic material, such as negative mass, it is unlikely we will develop a time machine capable of transporting humans in time. However, I do not want to underestimate humankind's ingenuity. NASA put a man on the moon using what we would view today as crude computers and space apparatus. It is entirely possible, to my mind, that someone will discover a unique engineering solution. Perhaps it will be Dr. Mallett. Perhaps it will be you.

Conclusions from Chapter 7

- The two greatest obstacles to time travel are:

    1. Understanding time to the point that it can be scientifically defined to the satisfaction of the scientific community.

    2. Efficiently creating significant quantities of energy capable of accelerating a spacecraft near the speed of light, or creating a traversable wormhole. This includes the creation of both positive and negative energy.

- The majority of the scientific community does not consider time travel paradoxes to be a significant obstacle to time travel. Numerous hypotheses have been proposed to resolve time travel paradoxes. Since we have already been able to cause subatomic particles to experience time dilation (i.e., forward time travel) without apparently causing a time travel paradox and disrupting the fabric of spacetime, this lends credence that time travel paradoxes are resolvable.

- Stephen Hawking's chronology protection conjecture, stating the laws of physics do not allow the appearance of closed timelike curves, is not widely accepted by the scientific community, and numerous physicists believe it is not valid. Even Dr. Hawking has acknowledged that some of the criticism is valid. My personal view is that unless the chronology protection conjecture is reformulated using a yet-to-be-developed-and-accepted quantum gravity theory, it must be held in abeyance. The jury remains out on its validity, but, like many of my colleagues, I do not think it makes a valid case that the laws of physics do not allow the appearance of closed timelike curves.

In this chapter, we have covered the greatest obstacles to time travel. Like all obstacles, unless forbidden by natural law, one day we will overcome them. This begs a question: When will we be able to time travel?

"The majority see the obstacles; the few see the objectives; history records the successes of the latter, while oblivion is the reward of the former."

—Alfred A. Montapert,
Author, *The Supreme Philosophy of Man: The Laws of Life*

CHAPTER 8

When Will We Be Able to Time Travel?

There is a saying in physics, "Everything not forbidden is compulsory." Many attribute this saying to American physicist and Nobel laureate Murray Gell-Mann, but actually, he borrowed it from T. H. White's 1958 novel, *The Once and Future King.*

The saying is widely used to suggest that in the fullness of time, science will achieve any goal not forbidden by physical law. Since no physical laws forbid human time travel, this suggests that eventually scientists and engineers will build a time machine capable of enabling human time travel. However, this begs a question. When will science or anyone build one?

Scientists, philosophers, and others argue no time machine exists yet to enable human time travel. This thinking generally relies on two beliefs:

1. We would be aware of time travelers.

2. Time travel will cause strange things to happen due to time travel paradoxes. For example, they argue that a person might just vanish because the person became a victim of the grandfather paradox.

Let us examine each belief and see if it holds up to logical reasoning.

First, if time travel exists and people travel in time, it will likely be the single most regulated technology ever invented by humankind. Because of its military potential, its science and engineering would be above top secret. The power required to enable human time travel would be enormous, and likely only be available to governments. If time travel paradoxes prove to be real, then the government regulation surrounding time travel would be greater. Unlike some science fiction novels suggest, time travel would not become a tourist industry. Relatives from the future would not casually drop by to say hello. Most likely, it would become the most closely guarded secret weapon of all time. For these reasons, it is reasonable to believe that real time machines would likely only be in the hands of governments, not individuals. It is also reasonable to believe that real time travelers would be highly trained government agents on above-top-secret missions. Given this as a likely reality, it would not be likely that we would know if time travelers from the future visit us, or are among us. They may already be here. There is anecdotal evidence, which we discussed in section 1.

Second, time travel paradoxes, as discussed in the last chapter, do not appear to present a real obstacle to time travel. To date, time dilation experiments, routinely performed in particle accelerators, do not appear to have any effect on the fabric of spacetime. We, of course, will not be able to conclusively prove the validity of time travel paradoxes until we have data from time travel experiments.

My point in presenting the above is a simple one. It is possible that a time travel machine exists or will exist in the future, and time travelers are visiting or will visit us. It is also possible, and even likely, the time travelers would keep their existence hidden for the reasons delineated above.

For the purposes of this chapter, let us assume a time machine capable of transporting humans in time does not exist. With this assumption, let us ask a deceptively simple question. When will a time machine capable of transporting humans in time be invented?

Mallett's prediction

Surprisingly, there are few prominent physicists addressing the above question. The most visible is Dr. Ronald Mallett from the University of Connecticut. He is currently working on an experiment to twist the fabric of spacetime using a laser. He claims to have found new solutions to Einstein's equations of general relativity, and he is basing his experiment on these solutions. According to Dr. Mallett, these new solutions suggest that using the light from a laser forced to traverse in a circle will twist spacetime, resulting in closed timelike curves (i.e., backward time travel). In an article (April 4, 2006) from phys.org, Dr. Mallett stated, "Einstein showed that mass and energy are the same thing. The time machine we've designed uses light in the form of circulating lasers to warp or loop time instead of using massive objects." From what I have read, I believe he is attempting to create a traversable wormhole. His plan is to send neutrons (i.e., a subatomic particle) into the circulating light beam and cause them to travel backward in time. If he is successful, and that is a big "if," he can actually send information back in time. How would this work? Subatomic particles, like a neutron, have a physical property called spin, which points either up or down. As an analogy, you can think of spin like a spinning top. For example, if you spin the top at one point, you can define it to be spin up. If you spin it upside down, then you can define it to be spin down. If Dr. Mallett sends a sequence of subatomic particles with appropriate spin, up or down, he would be able to transmit a binary code to the past. This would be an enormous accomplishment, if it works. To date, it does not work. However, Dr. Mallett predicts that he will get his time machine to work within a decade. As amazing as this accomplishment would be, it still would be a long way from transporting humans in time. It would, however, be a major leap in that direction.

The scientific community's prediction

After much research, I have concluded there is no solid consensus in the scientific community regarding the invention of a time machine capable of transporting humans in time. Many of the articles on time travel vaguely predict we should have a time machine sometime by the end of the century. However, this prediction is not the result of extrapolating from substantial scientific data. It appears more like a speculation. Such predictions are worthless, since they do not rest on scientific extrapolation, and one hundred years is so far out that few of us will be alive to remember if the article was right or wrong. I also believe this type of prediction violates a fundamental rule that futurists follow. The well-respected physicist Dr. Michio Kaku stated this rule in his popular book, *Physics of the Future*: "The key to understanding the future is to grasp the fundamental laws of nature and then apply them to the inventions, machines, and therapies that will redefine our civilization far into the future."

Stephen Hawking also believes that we will someday be able to travel in time. In fact, he predicts with almost certainty that we will be able to travel forward in time. He remains uncertain regarding traveling backward in time. According to an article on cnet.com (May 2, 2010), "Hawking: Time Travel Will Happen," Dr. Hawking posits, "At some point when we're all long gone, a day on a spaceship traveling at 650 million miles per hour would be akin to one year on Earth." While Stephen Hawking believes time travel to the future is both theoretically and practically possible, he gives no prediction as to when this will happen, and Hawking is not optimistic about time travel to the past.

Del Monte's prediction

I am going to base my predictions on Dr. Kaku's rule, namely, "grasp the fundamental laws of nature and then apply them to the inventions, machines, and therapies that will redefine our civilization far into the future."

Let us start with forward time travel. We are all traveling forward in time. Since none of us move anywhere near the speed of light, we are all traveling in time at roughly the same speed. Astronauts travel in space at relatively high speeds, and they theoretically experience time dilation. However, the speeds that today's spacecraft travel are only an insignificant fraction of the speed of light. Therefore, in practical terms, the astronauts do not experience significant time dilation. The amount of time dilation is small and, from the standpoint of the astronauts, an unnoticeable fraction of second. For example, at one point the Russian cosmonaut Sergei Avdeyev held the world record for time dilation experienced by a human being. In his three tours onboard the MIR spacecraft, cosmonaut Avdeyev orbited the Earth 11,968 times, traveling at approximately 17,000 miles per hour. As a result, Sergei Avdeyev is one fiftieth of a second younger than he would be if he had stayed on the Earth. One fiftieth of a second is measurable using sensitive scientific instruments, but to us humans, it is not noticeable.

Based on the published material regarding the current state of spacecraft propulsion systems, I predict we are getting close to prototyping a matter-antimatter propulsion system. In fact, if we have not already done it, I project we will in the near future, but the news probably will not be published. Here is the scientific data this projection is based on. First, we are able to make and store small quantities of antimatter using particle accelerators. I am aware that the published papers suggest that the process of making and storing antimatter is highly inefficient. According to the public record, the first creation of artificial antiprotons (i.e., antimatter) occurred in 1955. Since then, the production rate of antimatter increased nearly geometrically until the mid-1980s. In 1995, CERN produced a single antihydrogen atom and suspended it in a magnetic field. A simple Internet search will provide numerous plans to build a facility whose sole purpose is to create antimatter, and they argue that in time it will be economically feasible to create antimatter as a fuel. I recognize the Internet is not a reliable scientific source. However, the United States Air Force has been pursuing the production and storage of antimatter as a weapon since about 1960. It is reasonable to assume that the United States Air Force has spent a lot of money over the last half century. The exact amount, of course, would be top secret. However, given the weapons potential of an antimatter bomb, or the propulsion potential of an antimatter spacecraft engine, it is reasonable, in my opinion, to believe they have made more progress than what they publish on the NASA website or in scientific journals.

Let me summarize my predictions. First, we either have or will have a working prototype matter-antimatter propulsion engine in the near future. We will have a working matter-antimatter propulsion engine by 2050. Where does this number come from? Science is rapidly evolving, and scientific knowledge doubles about every ten years. In areas that have high focus and funding, scientific knowledge doubles about every five years. In my opinion, we are at the "Kitty Hawk" moment regarding the production, storage, and application of antimatter. About sixty years after the first sustained controlled airplane flight at Kitty Hawk, we were flying jet planes and had put a man on the moon. I believe the Kitty Hawk analogy fits the current-day production, storage, and application of antimatter extremely well. I leave it to you to make your own judgment on the applicability of the analogy.

Once we can routinely build matter-antimatter propulsion systems, spacecraft velocities will increase exponentially. I extrapolate that by the year 2075 we will have spacecraft powered by matter-antimatter propulsion systems, capable of velocities equal to about 650 million miles per hour. This may sound unbelievable, but compare how fast the Wright brothers' first plane flew to the speed of a 1960s-era jet plane. The Wright brothers' first plane flew to a speed of 6.8 miles per hour. By way of contrast, in 1961, the McDonnell Douglas F-4 Phantom II set a world speed record of 1,060 miles per hour. This increase in speed was achieved in less than sixty years. In about the same period, the early 1960s, the US/NASA North American X-15, a rocket-powered aircraft, set the official world record for the fastest speed by a manned aircraft, 4,520 miles per hour. Perhaps, understanding this analogy, achieving a speed of 650 million miles per hour over the next sixty years appears more reasonable.

At 650 million miles per hour, one day onboard the spacecraft will be equal to the passage of one year on Earth. Forward time travel will be a reality. By the end of this century, forward time travel will be akin to our current everyday air travel.

The prediction for backward time travel is not as clear cut. We have renowned physicists, such as Stephen Hawking, casting doubt that backward time travel is even possible without the use of exotic materials, such as negative mass.

While we can show via repeatable scientific experiments that the future can reach back into the past and influence experimental results (reference the "Twisting the arrow of time" section in chapter 1 and "The double-slit experiment" in chapter 6), we do not have any experiment that demonstrates time travel to the past.

In my opinion, the largest hurdle is energy. To date, adding positive energy, such as accelerating a particle, produces forward time travel (i.e., time dilation). There is no published literature suggesting we have been able to produce and store negative energy, or create negative mass.

To date, Dr. Mallett's spacetime twisting by light (STL) experiments have not been successful, and even Dr. Mallett is not predicting a short-term breakthrough. In 2006, Mallett predicted that time travel into the past would be possible within the twenty-first century. He went further and predicted that it may by possible within less than a decade, which would place us in the year 2016 (i.e., ten years from his prediction date). Obviously, Dr. Mallett is closer to his work than anyone else is. For a scientist of his stature to make such a prediction suggests that he is seeing encouraging results and has chosen not to publish them.

If, and when, anyone makes a breakthrough and is able to send a subatomic particle backward in time, it would just bring us to the point we are now with forward time travel. Therefore, I predict backward time travel for humans will not occur until close to the end of the twenty-first century.

Here is how I arrived at my prediction. If we assume Dr. Mallett, or anyone, is able to successfully cause a subatomic particle, like a neutron, to travel back in time before 2040, then transporting a human back in time becomes a similar problem to forward time travel. Using aviation as a model for the rate of technology development, I estimated that it would take about sixty years from where we are today with forward time travel (i.e., time dilation effects using subatomic particles) to reach a point where we could achieve human time travel to the future. Using similar logic, if, by the year 2040, we can achieve backward time travel using subatomic particles, then within sixty years (i.e., the end of the twenty-first century), we should be able to transport humans back in time. The pivotal point is achieving backward time travel for subatomic particles. When that occurs, I expect progress will be at a rate similar to forward time travel.

Conclusions from Chapter 8

- It is possible that a time machine capable of transporting humans in time already exists. Because of its military potential, and the amount and kind of energy a time machine would use, it is likely in the hands of governments, not individuals. As such, time travel would be the most closely guarded secret of all time, and highly regulated by the governments with time travel capability.

- It is possible that time travelers are already visiting us. Such visitors would likely be highly trained government agents and would know how to avoid detection.

- If time travel paradoxes turn out to be real, time travel would be considered extremely risky and be even more highly regulated.

- If we assume a human-capable time travel machine does not exist, we are close to developing one and may have a prototype for forward time travel by 2050 (i.e., the mid-twenty-first century).

- Backward time travel will require a more complex time machine and potentially exotic fuels, such as negative mass. It is unlikely we will have a working prototype for backward human time travel until the end of the twenty-first century.

Many elements of this chapter appear to have roots in what the Department of Defense calls "black projects." Such projects are a highly classified military/defense projects, unacknowledged publicly by the government, military personnel, and defense contractors. Projects dealing with time travel will likely be black projects. Time travel itself has immense potential as a weapon. The technologies used to enable time travel, such as antimatter and negative energy, would also have immense potential as weapons. Therefore, there is little doubt that time travel projects would be black projects and classified, in the vernacular, "above top secret." The formal term for these types of projects in the United States is special access programs (SAPs). However, it is reasonable to ask, why all secrecy? We are not going to get a complete look behind the curtain, but there is evidence that time travel is real. We discussed the evidence in the first section of this book. This may answer the question, "Why all the secrecy?"

"I am become death, the destroyer of worlds."

—J. Robert Oppenheimer,
Reflections after witnessing the first atomic bomb detonation on July 16, 1945, at the Trinity test in New Mexico

CHAPTER 9

Preserve the World Line

At this point, it is evident that the question is not "if" we will develop a time machine capable of enabling human time travel, but "when." It is highly probable that human time travel will be achieved by the end of the twenty-first century.

Time travel will be the ultimate weapon. With it, any nation can write its own history, assure its dominance, and rule the world. However, having the ultimate weapon also carries the ultimate responsibility. How it is used will determine the fate of humankind. These are not just idle words. This world, our Earth, is doomed to end. Our sun will eventually die in about five billion years. Even if we travel to another Earth-like planet light-years away, humankind is doomed. The universe grows colder each second as the galaxies accelerate away from one another faster than the speed of light. The temperature in space, away from heat sources like our sun, is only about 3 degrees Kelvin (water freezes at 273 Kelvin) due to the remnant heat of the big bang, known as the cosmic microwave background. As the universe's acceleration expands, eventually the cosmic microwave background will disperse, and the temperature of the universe will approach absolute zero (-273 degrees Kelvin). Our galaxy, and all those in our universe, will eventually succumb to the entropy apocalypse (i.e., "heat death") in a universe that has become barren and cold. If there is any hope, it lies in the technologies of time travel. Will we need to use a traversable wormhole to travel to a new (parallel) universe? Will we need to use a matter-antimatter spacecraft to be able to traverse beyond this universe to another?

I believe the fate of humankind and the existence of the universe are more fragile than most of us think. If the secrets of time travel are acquired by more than one nation, then writing history will become a war between nations. The fabric of spacetime itself may become compromised, hastening doomsday. Would it be possible to rip the fabric of spacetime beyond a point that the arrow of time becomes so twisted that time itself is no longer viable? I do not write these words to spin a scary ghost story. To my mind, these are real dangers. Controlling nuclear weapons has proved difficult, but to date humankind has succeeded. Since Fat Man, the last atomic bomb of World War II, was detonated above the city of Nagasaki, there has been no nuclear weapon detonated in anger. It became obvious, as nations like the former Soviet Union acquired nuclear weapons, that a nuclear exchange would have no winners. The phrase "nuclear deterrence" became military doctrine. No nation dared use its nuclear weapons for fear of reprisal and total annihilation.

What about time travel? It is the ultimate weapon, and we do not know the consequences regarding its application. To most of humankind, time travel is not a weapon. It is thought of a just another scientific frontier. However, once we cross the time border, there may be no return, no do-over. The first human time travel event may be our last. We have no idea of the real consequences that may ensue.

Rarely does regulation keep pace with technology. The Internet is an example of technology that outpaced the legal system by years. It is still largely a gray area. If time travel is allowed to outpace regulation, we will have a situation akin to a lighted match in a room filled with gasoline. Just one wrong move and the world as we know it may be lost forever. Regulating time travel ahead of enabling time travel is essential. Time travel represents humankind's most challenging technology, from every viewpoint imaginable.

I have considered this seriously. What regulations are necessary? I have concluded they need to be simple, like the nuclear deterrence rule (about thirteen words), and not like the US tax code (five million words). When you think about it, the rule of nuclear deterrence is simple: "If you use nuclear weapons against us, we will retaliate, assuring mutual destruction." That one simple rule has kept World War III from happening. Is there a similar simple rule for time travel?

I think there is one commonsense rule regarding time travel that would assure greater safety for all involved parties. I term the rule "preserve the world line." Why this one simple rule?

Altering the world line (i.e., the path that all reality takes in four-dimensional spacetime) may lead to ruination. We have no idea what changes might result if the world line is disrupted, and the consequences could be serious, even disastrous. The preserve the world line rule has two implications:

1. Initially, we should travel only to the near past or near future (such as a few seconds into the past or future), and only for a short duration (such as a few seconds). This appears, to my mind, long enough to witness an event, such as the finish of a horse race, but short enough to minimize doing something that

could seriously alter the past or the future. As we learn more about time travel, we will understand how deeply we can travel into the past or future, and what durations would be advisable, without suffering the consequences of an ill-fated world line.

2. Any time traveler should only observe, not act. This is similar to the noninterference directive in the popular *Star Trek* television series and films. It is also imperative that we do not use knowledge gained from the future, such as the winner of a horse race, to alter our actions, such as making a large wager on the known winner of a horse race. Observation is likely not to cause a disruption in the world line, but it would be sufficient for scientific purposes.

The preserve the world line rule is akin to avoiding the "butterfly effect." This phrase was popularized in the 2004 film *The Butterfly Effect*, with the now famous line: *"It has been said that something as small as the flutter of a butterfly's wing can ultimately cause a typhoon halfway around the world."* Although the line is from a fictional film, the science behind it is chaos theory, which asserts there is a sensitive dependence on the initial conditions of a system that could result in a significant change in the system's future state. Edward Lorenz, American mathematician, meteorologist, and a pioneer of chaos theory, coined the phrase "butterfly effect." For example, the average global temperature has risen about one degree Fahrenheit during the last one hundred years. This small one-degree change has caused the sea levels around the world to rise about one foot during the same period. Therefore, I believe, it is imperative not to make even a minor change to the past or future during time travel until we understand the implications.

If it is possible to adhere to the preserve the world line rule, traveling in time may become a safe experiment. Remember, our first nuclear weapons were small compared to today's nuclear weapons. Even though they were comparatively small, the long-term effects included a 25% increase in the cancer rate of survivors during their lifetime. We had no idea that this side effect would result. Similarly, we have no idea what the long-term effects will be if we alter the world line. We already know from laboratory experiments that the arrow of time can be twisted. Things done in the future can alter the past. Obviously, altering the past may alter the future. We do not know much about it because we have not time traveled in any significant way. Until we do, preserving the world line makes complete sense.

Conclusions from Chapter 9

- Time travel has the potential to destroy the fabric of spacetime itself.

- We need to follow the "preserve the world line" rule:

  o Keep initial time travel experiments limited to short durations into the near past and near future.

  o Observe, do not act, during or after time travel.

Parting Thoughts from the Author

I hope you enjoyed reading *How to Time Travel* as much as I enjoyed writing it.

My objective in writing this book was to address what appears to be the most popular question in science: how to time travel. I attempted to make the science easy to understand, so that anyone with a high school or higher education would be able to grasp the essence of the scientific principles that govern time travel. I also sought to keep the mathematics to a minimum, but provided significant mathematical detail in appendices 1, 2, and 3 for those who choose to go further into the mathematics governing time travel. I want this book to be enjoyable for every reader. I also attempted not to bias the time travel data and scientific information presented throughout the book. I thought it best that you make your own judgments.

This book is not simply a rehashing of previously published or available material. Several new items are original works. These include:

- The existence equation conjecture (chapter 4, appendices 2 and 3)

- The grandchild paradox (chapter 6)

- The preserve the world line rule (chapter 9)

- The time uncertainty interval (appendix 5)

I sincerely hope our world lines cross and we meet in person, even if that meeting requires time travel. Until then, let me leave you with this one thought. Next time someone asks what time it is, do not look at your watch, but rather look around you. If everything seems normal, you can reply, "It is the present."

Glossary

Absolute zero—The theoretical temperature at which all motion stops, even at the atomic level, and the substance has no heat energy. According to the laws of physics, it is not possible for a substance to reach absolute zero.

Accelerating expansion of the universe—It is a scientific fact that the expansion of the universe is accelerating. The farther away a galaxy is from us, the faster it is accelerating away from us. In fact, the acceleration of the farthest-away galaxies appears to exceed the speed of light. Modern science believes that it is space that is expanding faster than the speed of light, which makes it appear that the galaxies themselves are accelerating faster than the speed of light. No law of physics prevents space from expanding faster than the speed of light.

Acceleration—The change in velocity (speed) of an object over a specific time.

Advanced aliens—This refers to intelligent extraterrestrial life that is scientifically knowledgeable and able to apply technology, such as radio broadcasting.

Anthropic principle—This principle asserts the universe is the way it is because if it were different, we would not exist.

Anthropogenic warming—This is global warming caused or influenced by humans.

Antimatter—The mirror image of matter. In matter, the electrons in atoms are negative, and the nucleus is positive. In antimatter, the electrons in atoms are positive, and the nucleus is negative.

Baryogenesis theories—A class of theories that asserts the existence of a process that creates an asymmetry between matter and antimatter in the universe. Baryogenesis theories are postulated to explain the absence of antimatter in the universe. These theories are unproven and considered speculative.

Big bang duality—This theory postulates that the big bang did not originate as a singularity (one highly dense energy particle), but a duality (one highly dense energy particle and one highly dense energy antiparticle). The expansion of energy (commonly referred to as the big bang), according to the big bang duality theory, was the result of a matter-antimatter particle collision in the "bulk."

Big bang theory—The theory that the universe originated from a highly dense energy state that expanded to form all that we observe as reality.

Big crunch—This theory held that gravity would eventually overcome the expansion of the universe and force all reality into a highly dense energy point.

Black hole—A region of space whose gravitation is so intense that not even light can escape, thus the name "black hole."

Bubble universes—Entire universes that theoretically exist in the "bulk." The bubble universes may or may not resemble our own.

Bulk—A superuniverse capable of holding countless universes. In theory, it contains our own universe, as well as other universes.

Casimir-Polder force—The attractive force between two parallel metal plates placed extremely close together (approximately a molecular distance) in a vacuum. Science believes the attraction is due to a reduction in virtual particle formation between the plates. This, in effect, results in more virtual particles outside the plates, whose pressure pushes them together.

Celsius—A temperature scale in which water freezes at zero degrees (0°C) and boils at one hundred degrees (100°C).

Chaotic inflation theory—This theory asserts that the universe does not inflate uniformly, but may accelerate in regions of space devoid of matter and radiation. The nonuniform inflation causes portions of the universe to separate from the existing universe. These portions form miniuniverses ("bubble" universes). The word "bubble" is used to describe this concept because when bubbles form, occasionally a smaller bubble will form on a larger bubble, and then separate from it.

Chronology protection conjecture—This is a conjecture by physicist Stephen Hawking, published by him in a 1992 paper, which asserts the laws of physics do not allow the appearance of closed timelike curves. This conjecture has been a subject of debate and has substantial supporters and adversaries. We have no consensus that it is a scientific fact.

Classical mechanics—Refers to Newton's three laws of motion, enunciated by Newton in the seventeenth century. It is widely used today by almost everyone since it provides excellent agreement with phenomena in our everyday experiences (the macroworld). For example, if you play billiards, Ping-Pong, or even marbles, you are intuitively using Newton's laws of motion. Newtonian mechanics and Newton's laws of motion are synonymous with the term "classical mechanics."

Closed timelike curves—In mathematical physics, a closed timelike curve is a solution to the equations of general relativity that demonstrates a particle's world line closes on itself (returns to the starting point). Numerous theoretical physicists interpret this to imply time travel to the past is possible.

Conservation of energy—Arguably the most sacred law in physics, namely, that energy cannot be created or destroyed, only transformed from one form to another.

Conservation of mass—This law is similar to the conservation of energy, namely, mass cannot be created or destroyed, only transformed from one form to another, including the transformation to energy.

Cosmological constant—An arbitrary constant, originally proposed by Albert Einstein, to force his equations of general relativity to predict a static universe. In 1929, when Edwin Hubble discovered the universe was expanding, Einstein proclaimed the cosmological constant his "greatest blunder." Today, physicists are using the cosmological constant, along with mathematical manipulation of general relativity, to model the accelerated expansion of the universe.

Cosmological horizon—The observable universe. Its basis is the time light has had to travel to us from the edge of the observable universe, in accordance with the age of the universe (13.7 billion years old).

Dark energy—A theoretical force postulated to be responsible for the accelerated expansion of the universe. Some scientists describe it as a "vacuum" force, while others describe it as a "negative" force. In reality, no scientific consensus exists regarding the nature of dark energy, including its existence.

Dark matter—This is mass in galaxies, galaxy clusters, and between galaxy clusters that cannot be observed directly, but is postulated to exist to explain why the farthest-away stars from the center of the galaxy rotate at approximately the same rate as the stars nearer the center of the galaxy. From the laws of physics, the farthest-away stars should be rotating much slower than those nearer the center. Therefore, scientists have postulated that more mass, namely, dark matter, is in the galaxy than we observe, since it did not emit or reflect light. The existence of dark matter has been confirmed by gravitational lensing (gravity bending light) observations. Scientific calculations indicate that dark matter may account for about 90% of the total matter of the universe. Scientific speculation asserts that a particle is associated with dark matter, namely, the WIMP particle (Weakly

Interacting Massive Particle). To date, there is no conclusive evidence that the WIMP particle exists.

Del Monte paradox—An observation made by Louis Del Monte and introduced in his book Unraveling the Universe's Mysteries, namely, each significant scientific discovery results in at least one profound scientific mystery.

Dilated—Typically used as a verb in this book, this term refers to one clock traveling at the speed of light running slower than another clock at rest.

Dirac sea—A theory postulated in 1930 by Paul Dirac (a British physicist) that empty space (a vacuum) consists of a sea of virtual electron-positron pairs. This eventually led to the discovery of antimatter.

Doppler shift—The elongation or compression of a wavelength of light or sound, which depends on the motion of the emission source relative to an observer. For example, light's wavelength elongates as the galaxies farthest from Earth move away from us due to the expansion of space. The Doppler shift is toward the red portion of the spectrum, since red is the color of longer wavelengths of light.

Earth-like—Refers to a planet that is similar to Earth. Typically, the planet would be in the habitable zone of the star it orbits, be large enough for gravity to hold its atmosphere in place, and have liquid water, to name a few of the properties of an Earth-like planet.

Entropy—In nature, all processes produce waste energy (heat) that escapes and becomes unavailable to do mechanical work. The waste energy that escapes is a measure of the entropy and increases the disorder of a system. Since the universe is continually evolving through cosmic processes, the entropy of the universe is always increasing.

Entropy apocalypse (a.k.a. "heat death")—A point in time when all energy is heat, and unavailable to do mechanical work. It spells doom for all life.

Euclidean geometry—The geometry developed by the ancient Greek mathematician Euclid. Mathematicians refer to it as a "flat" geometry, where parallel lines remain equidistant (the same distance apart) to infinity, and the sum of angles within a triangle equals 180 degrees. We typically learn this geometry as schoolchildren.

Existence equation conjecture—This theoretical equation enables the calculation of a mass's kinetic energy as it moves in the fourth dimension of Minkowski space. This equation is unusual in that it indicates the kinetic energy of a mass moving in the fourth dimension is negative, and massively large. A speculative interpretation of the existence equation conjecture is that existence requires negative energy, which is being siphoned from the universe, and results in the accelerated expansion of the universe. The existence equation conjecture is discussed more fully in chapter 4, appendix 2, and appendix 3.

Existential risk—Any risk with the potential to destroy humankind or drastically restrict human civilization. In theory, an existential risk could end the existence of Earth, the solar system, the galaxy, or even the universe.

Extraterrestrial life—Life that exists outside of the Earth. For example, if we discover microbial life on Mars, it would be proof of extraterrestrial life.

Fahrenheit—A temperature scale at which water freezes at thirty-two degrees (32°F) and boils at two hundred twelve degrees (212°F).

Field—A concept in the physical sciences used to imply mediation at any distance, without the use of particles as mediators. For example, gravity exerts an attractive force between two masses via a gravitational field. If, however, science discovers the particle of gravity called the graviton, the mediation (the force-carrying particles) between the two masses will not be a field, but will be by graviton particles. The field concept in physics is an old concept and predates understanding the role that particles play as mediators.

Fossil record—Generally refers to the total number of fossils found, and the information derived from them by paleontologists, which are scientists who study fossil (prehistoric) animals and plants.

Four fundamental forces—Modern physics recognizes four known fundamental forces, namely, gravitational, electromagnetic, strong nuclear, and weak nuclear. All physical forces theoretically trace back to one or more of these forces.

Fourth dimension—In a 1912 manuscript on relativity, Einstein equated the fourth dimension to ict (where $i = \sqrt{-1}$, c is the speed of light in empty space, and t is time, representing the numerical order of physical events measured with "clocks"). The entire thrust of using four-dimensional space in the special theory of relativity is attributed to Russian mathematician Hermann Minkowski. In 1907, Minkowski demonstrated Einstein's special theory of relativity (1905), presented algebraically by Einstein, could be presented geometrically as a theory of four-dimensional spacetime.

Galaxy—A system of millions to billions of stars, along with gas and dust, held together by gravity. Most stars, like our sun, have planets and other celestial bodies orbiting them. There are billions of galaxies in our universe. Typically, galaxies are separated from one another by vast regions of space (measured in light-years).

Gamma rays—Short-wavelength, high-energy, electromagnetic radiation emitted by radioactive substances.

General theory of relativity—A theory developed by Albert Einstein dealing with gravity and noninertial frames of reference. It is termed the "general theory of relativity" to differentiate it from the "special theory of relativity," which focuses on inertial frames of reference.

Grandchild paradox—Any situation, real or imagined, that reverses the "arrow of time" and allows the future to influence the past.

Grandfather paradox—A situation where a person travels back in time and prevents his grandfather from meeting his grandmother. This raises the question, "What happens to the person (i.e., the grand-child)?"

Gravitational lensing—The phenomenon that light's path can be affected by gravity. For example, light from distant galaxies bends as it passes through the gravitational field of another galaxy. This can result in magnifying, distorting, or producing multiple images of the original light source for a distant observer.

Gravity (or gravitation)—The attractive force one mass exerts on another mass.

Ground state—The lowest energy state of an atom or particle.

Ground-state entropy—The lowest entropy state of a system.

Heisenberg uncertainty principle—This scientific principle holds it is impossible to accurately determine both the position and velocity of a particle simultaneously. In the context of quantum mechanics, it conceptually conveys the probabilistic nature of physical phenomena at the atomic and subatomic level (quantum level).

Hubble volume—The region of the universe surrounding an observer, beyond which objects recede from the observer at a rate greater than the speed of light due to the expansion of the universe.

Inertial frame of reference—A frame of reference at rest or moving at a constant velocity.

Kelvin—The International System of Units (SI) absolute thermodynamic temperature scale, using as its null point zero degrees Kelvin (0°K), which refers to a state devoid of all heat and motion. The Kelvin temperature scale equates to the Celsius temperature scale via the following equation: K = [°C] + 273.15.

Kinetic energy—The energy associated with an object due to its motion.

Lamb shift—This refers to the small difference in energy between two states of the hydrogen atom. American physicist Willis Eugene Lamb (1913–2008) first detected it and received the Nobel Prize in Physics in 1955 for his discoveries related to the Lamb shift.

Macrolevel—This is our everyday world. It is the reality that we can typically see and touch.

Mass—In physics, this typically refers to matter, such as a subatomic particle, atom, or an assembly of subatomic particles and atoms.

Mediators—In physics, mediators are the particles that carry force between entities. For example, the photon is the force carrier for the electromagnetic force.

Minkowski space—The mathematical, four-dimensional concept of spacetime developed by Hermann Minkowski in his 1909 paper "Space and Time." Minkowski spacetime found application in Einstein's special theory of relativity and in the development of the existence equation conjecture.

Miracle—An act by a supernatural being that typically suspends the laws of physics.

Moore's law—This is more of a general rule or observation than an actual physical law. Moore's law states the number of transistors that can be placed inexpensively on an integrated circuit doubles approximately every two years.

Multiverse—The concept that there are other universes beyond our own. The phrases "parallel universes," "alternative universes," "quantum universes," "parallel dimensions," and "parallel worlds" are synonymous to the multiverse.

Muon—A negatively charged particle approximately two hundred times more massive than an electron.

Neutrino—An elementary particle with close to zero mass and no electrical charge, and that travels close to the speed of light.

Newton's laws of motion—See classical mechanics.

Newtonian mechanics—See classical mechanics.

Nomads—Groups of people, typically a clan or tribe, whose civilization survives by moving from one location to another location more favorable to their survival.

Occam's razor—A principle of science that holds the simplest explanation is the most plausible one, until new data to the contrary become available.

Particle—In physics, this can refer to a massless object or, alternately, a small object with mass. A photon is an example of a massless object. A muon is an example of a small object with mass.

Particle accelerator—An apparatus that uses electromagnetic fields to accelerate subatomic particles to high velocities, even velocities approaching the speed of light.

Photoelectric effect—The ejection of electrons from any substance due to the incidence of electromagnetic energy (light).

Photon—A particle (energy packet) of light (electromagnetic radiation). The photon has no mass and travels at the speed of light in a vacuum.

Planck length—The smallest unit of length theoretically possible, which suggests that space itself may be quantized. The Planck length is equal to approximately 1.6 x 10-35 meters, and is defined using three fundamental physical constants: the speed of light in a vacuum, Planck's constant, and the gravitational constant. At the Planck length, gravity is thought to become as strong as the three other fundamental forces (electromagnetic, strong nuclear, and weak nuclear), and quantum effects dominate.

Planck time—The time it takes light in a vacuum to travel one Planck length. This is the smallest unit of time in which science believes change can occur. It is approximately equal to 10-43 seconds. This implies that time itself may be quantized.

Products—The resulting substances of a chemical reaction. The substances may be compounds, elements, or both.

Quantized—The discrete nature of a substance, like mass or energy. In this book, it is typically used as a verb when describing the discrete nature of reality (mass, energy, space, and time).

Quanta—A discrete packet of energy, such as a photon.

Quantum—Synonymous with quanta.

Quantum entanglement—A phenomenon in quantum mechanics where a pair of particles or photons interacts with each other and forms an invisible bond. When a pair of particles becomes entangled, their quantum states, which completely describes their state of being, communicate and correlate with each other, even when the particles are separated by a distance. Thus, changing the quantum state of one entangled particle forces the quantum state of the other entangled particle to change in a way that they remain in a correlated quantum harmony. This phenomenon is a scientific fact, but not completely understood. One mystery is that the communication between entangled particles appears to travel faster than the speed of light in a vacuum.

Quantum fluctuation—An effect observed in quantum physics in which a temporary change in the amount of energy occurs at a point in space, in accordance with the Heisenberg uncertainty principle. This effect gives rise to "virtual particles" or "spontaneous creation." This is a scientific fact.

Quantum level—The scale of atoms and subatomic particles.

Quantum mechanics—The scientific principles that explain the behavior of physical objects and their interactions with energy on the scale of atoms and subatomic particles.

Quantum universe—A theory that the entire universe consists of quantized matter and energy.

Quark—The quark is an elementary particle and a fundamental building block of other particles, like protons and neutrons. There are six types of quarks, known as flavors, including up, down, strange, charm, bottom, and top.

Qubit—In quantum computing, the qubit is the quantum bit of information analogous to the classical computer bit. Whereas the classical computer bit contains information and can represent a one or zero, the qubit can represent a one, zero, and a superposition of both at the same time. For example, using a classical computer bit, a polarized photon could be expressed by either a one for horizontal polarization or a zero for vertical polarization. A cubit represents both states simultaneously.

Reactants—Refers to the substances (elements and compounds) that react in a chemical reaction to form one or more new substances (products).

Redshift—Refers to the elongation of a light wave as its wavelength stretches, due to the emission source moving away from an observer. Longer wavelengths of light are in the red portion of the spectrum.

Relativistic mechanics—This refers to any form of mechanics that is derived from and/or compatible with Einstein's general and special theory of relativity.

Singularity—In the context of the big bang, this refers to the point of infinitely dense energy that gave birth to the big bang.

Spacetime—This refers to the concept that time is dependent on space. In effect, space and time are fused together in a mathematical model to form a continuum known as spacetime. This has enabled physicists to simplify numerous physical theories and describe the universe more precisely.

Special theory of relativity—A theory developed by Albert Einstein, based on two postulates:

1. Physical laws have the same mathematical form in any inertial system (a system at rest or moving at a constant velocity).

2. The velocity of light is independent of the motion of its source, and will have the same value when measured by observers moving with constant velocity with respect to each other.

From these two fundamental postulates, Einstein was able to develop his famous mass-energy equivalence equation, the time dilation equation, and the relativistic kinetic energy equation. The special theory of relativity is one of the most successful theories of modern science.

Speed of light in a vacuum—This is the speed at which light (electromagnetic radiation) travels in a vacuum, which is exactly 299,792,458 meters/second. Scientists universally view light as the upper speed limit in the universe. Nothing travels faster in a vacuum than the speed of light.

Spiral galaxy—A galaxy having a spiral form with spiral arms. The oldest stars in a spiral arm are near the center of the galaxy. Our Milky Way galaxy is a spiral galaxy.

Spontaneous symmetry breaking—A theory that holds a system in a symmetrical state is able to transform to an asymmetrical state.

Standard model—This refers to the standard model of particle physics, which mathematically models the behavior of elementary particles and their interaction relative to the electromagnetic, strong nuclear, and weak nuclear forces.

Star—This is a self-luminous celestial body, like our sun, consisting of gas held together by gravity. The luminescence is the result of nuclear reactions within the body whose energy makes its way to the surface and emits as radiation.

String theory—A mathematical theory that represents all matter, such as subatomic particles, as consisting of strings that vibrate in one dimension and exist in eleven dimensions. A number of prominent physicists consider string theory to be a contender for the theory of everything.

Supernatural being—A being that exists outside the natural realm. The words "deity" and "god" are synonymous with a supernatural being. Depending on specific religious beliefs, the supernatural being has specific powers over the natural world.

Superuniverse—See bulk.

Symmetry of physical laws—A concept in physics that argues that a physical law is unchanged by any theoretical transformation.

This is a simple example: a geometric sphere maintains all elements that define it as a geometric sphere, regardless of any rotational transformation. Most physicists believe physical laws are symmetrical.

Theory of everything—A self-contained mathematical model that describes all fundamental forces and forms of matter.

Thought experiment—This is a conceptual experiment. It considers a hypothesis, theory, or principle and thinks through the ramifications to illustrate a point. A thought experiment may or may not be possible to perform in reality. The objective is to explore the hypothesis, theory, or principle and its potential consequences. Einstein is historically considered the master of the thought experiment, or gedankenexperiment (in German).

Time—The traditional definition of time is the numerical sequence of events as measured by clocks.

Time dilation—A phenomenon predicted by Einstein's special and general theories of relativity. A clock moving close to the speed of light, or placed in a strong gravitational field, will record time slower than a clock at rest or in a weak gravitational field. This effect is independent of the clock mechanism.

Twin paradox—A situation where a twin boards a spacecraft and travels at close to the speed of light. Because of time dilation, the twin on the spacecraft ages much slower than the twin who remained on Earth.

Uncertainty principle—See the "Heisenberg uncertainty principle."

Velocity—The distance an object travels divided by the time it takes to travel that distance.

Virtual particle—A particle that exists for a limited time and obeys some of the laws of real particles, including the Heisenberg uncertainty principle and the conservation of energy. However, its kinetic energy may be negative.

Wavefunction—The wavefunction, in quantum mechanics, describes the probability of a particle's state (position, momentum, and other attributes).

Wave-particle duality—Refers to the exhibition of both wave and particle properties.

World line—The unique path that an object takes as it travels through four-dimensional spacetime.

APPENDIX 1

Time Dilation Equations

Time dilation due to kinetic energy

According to special relativity, as a mass—for example, a clock—moves close to the speed of light, time slows down relative to a clock at rest. To understand the equation to calculate time dilation, consider a clock at rest. Let us define time between "ticks" for this clock as Δt_0. Next, consider a clock moving close to the speed of light. Let us define the ticks for this clock as Δt. The relationship between Δt and Δt_0 is:

$$\Delta t = \Delta t_0 \frac{1}{\sqrt{1 - v^2/c^2}}$$

Where v^2 is the speed of motion containing the clock whose time interval between ticks in that frame is Δt, as observed from a frame of reference at rest containing a clock whose time interval between ticks is Δt_0.

This equation's derivation comes from the special theory of relativity. Please note that the mathematical expression $\sqrt{1 - v^2/c^2}$ is going to be between 0 and 1, for values of v between 0 and c (i.e., the speed of light).

Let us do a simple calculation to illustrate how the equation works. Assume you are on a spacecraft that can travel close to the speed of light, for example, .95 times the speed of light (c), or 95c. If you spend ten years traveling at this velocity before returning to Earth, then 32 years will have passed on Earth. This can be verified by setting $v = .95c$, $\Delta t_0 = 10$ years, and solving for Δt. This demonstrates that time dilation is equivalent to forward time travel.

Time dilation due to gravity

Next, let us consider the formula for gravitational time dilation. As we discussed earlier in the book, a clock in a strong gravitational field moves slower. Here is the formula for gravitational time dilation due to a large mass:

$$t_d = t \sqrt{1 - \frac{2Gm}{c^2 r}}$$

Where:

- t_d = time inside the gravitational field

- t = time outside the gravitational field

- m = the mass causing the gravitational field

- r = the distance from the center of the gravitational field

- c = the speed of light in a vacuum.

- G = the universal gravitational constant (6.6742 X 10^{-11} N m^2 kg^{-2}), originally calculated from Newton's law of gravitation. It is an empirical physical constant involved in the calculation(s) of the gravitational force between two bodies.

Notice, the gravitational constant (G) is an extremely small fraction. To observe time dilation due to a mass m, the mass must be large. However, with the advent of atomic clocks, even minor time dilations effects are now measurable.

If the gravitational time dilation is due to an accelerated frame of reference along a straight line, the formula for time dilation becomes:

$$T_d = e^{\frac{gh}{c^2}}$$

Where:

- T_d = the *total* time dilation at a distant position

- g = the acceleration of the box as measured by the base observer

- h = the "vertical" distance between the observers

- c = the speed of light

- e = is an important constant, approximately equal to 2.71828, and serves as the base of the natural logarithm system of mathematics

If the gravitational time dilation is due to a rotating disk, where the observer is located at the center of the disk and co-rotating with it, the equation becomes:

$$T_d = \sqrt{1 - r^2 w^2 / c^2}$$

Where:

- r = the distance from the center of the disk (which is the location of the observer)

- w = the angular velocity of the disk

APPENDIX 2

Derivation of the Existence Equation Conjecture

From the special theory of relativity, we know that the relativistic kinetic energy may be calculated using the following equation:

$$E_k = \frac{mc^2}{\sqrt{1 - \frac{v^2}{c^2}}} - mc^2$$

Where E_k is the relativistic kinetic energy, m is the rest mass of an object, v is the velocity of an object, and c is the speed of light in a vacuum.

It is important to note that in the derivation of the relativistic kinetic energy, m is the rest mass of an object, since we will use the equation to derive the existence equation conjecture.

In the special theory of relativity, Einstein used Minkowski's four-dimensional spacetime, whose vectors are defined by four coordinates, X_1, X_2, X_3, and X_4. Note, X_1, X_2, and X_3 are the typical coordinates of the three-dimensional space, and $X_4 = ict$, where $i = \sqrt{-1}$, c is the speed of light in empty space, and t is time, representing the numerical order of physical events measured with clocks. (The mathematical expression i is termed an imaginary number because it is not possible to solve for the square root of a negative number.) $X_4 = ict$ is a spatial coordinate, and is on equal footing with X_1, X_2, and X_3 (the typical coordinates of three-dimensional space). X_4 is not a "temporal coordinate."

If we assume the object is at rest, and X_1, X_2, and X_3 are each equal to zero (i.e., the object is at the origin of the coordinate system), it follows that the velocity associated with them is zero. However, the object continues to move in time, since the X_4 coordinate contains a time component. In addition, we will assume that coordinate time at an event is identical to the proper time, which implies the observer and clock are at the same location as the event, and the clock at the event is synchronized to the observer's clock. Given these conditions, we use the discipline of calculus to differentiate the Minkowski vector $(0,0,0, ict)$ with regard to time (to get the velocity associated with this vector). The result is a scalar quantity and represents the velocity associated with moving along the X_4 axis. Using this methodology, the velocity v_4 along the X_4 is equal to ic. Using this result and the relativistic kinetic energy equation given above, we can calculate the associated relativistic kinetic energy E_{kx4}. The result is:

$$E_{kx4} = - .3mc^2$$

Where m is the rest mass of the object, and c is the speed of light in a vacuum. Note, for simplicity, the equation is rounded to one significant digit after the decimal point.

The equation ($E_{kx4} = - .3mc^2$) is termed the existence equation conjecture. To a first order, the existence equation conjecture appears to agree with experimental time dilation results (see appendix 3). Until we have more data, and the scientific community weighs in on its validity, I have labeled the equation a conjecture, hence its name: existence equation conjecture.

APPENDIX 3

Experimental Verification of the Existence Equation Conjecture

From particle accelerator data (Bailey, J., et al., *Nature* 268, 301 [1977] on muon lifetimes and time dilation): in an experiment at CERN by Bailey et al., muons of velocity $0.9994c$ were found to have a lifetime 29.3 times the laboratory lifetime.

If a muon needs $KE_{X4} = -.3mc^2$ at rest for one lifetime, per the existence equation conjecture, conceptually it requires the following negative energy to exist for 29.3 lifetimes:

$$E_{X4} = 29.3 \ x -.3mc^2$$

(In effect, this argues the amount of negative energy required for existence increases. This suggests that an enormous amount of positive energy would be needed to cancel the negative energy of existence.)

If we do the mathematics, we get the following result:

$$E_{X4} = -8.8mc^2$$

We must judge that the kinetic energy is completely consumed when the particle ceases to exist (decays) as a muon. (We know that muons decay to one electron and two neutrinos, and sometimes the decay produces other particles, such as photons.) Please note, in keeping with the accuracy of the lifetime data (observed time dilation), we will round off all calculations of kinetic energy to one decimal point.

We will consider the Bailey experiment, and calculate the relativistic kinetic energy of a muon traveling at .9994c:

Relativistic calculation of the muon's kinetic energy from Einstein's relativistic kinetic energy equation, delineated below:

$$E_k = \frac{mc^2}{\sqrt{1 - \frac{v^2}{c^2}}} - mc^2$$

If we let $v = .9994c$, $v^2 = .9988c$, reducing to two significant digits (not rounding up to $1.0c$ to avoid conflict with the theory and an irrational result) implies $v^2 \sim .99c^2$

Then,

$$E_k = \frac{mc^2}{\sqrt{.01}} - mc^2$$

$$E_k = \frac{mc^2}{.10} - mc^2$$

$$E_k = 10 \, mc^2 - mc^2$$

$$E_k = 9.0mc^2$$

Therefore, the kinetic energy of a muon accelerated to approximately 99.9% the speed of light is approximately $9.0mc^2$.

Discussion of Results:

Theoretically, the existence equation conjecture predicts – 8.8 mc^2 of negative energy is required for a 29.3 lifetime extension (the time dilation we observe).

The experimental kinetic energy (one decimal point accuracy) is calculated to be 9.0 mc^2. This positive kinetic energy completely balances the negative energy within ~ 2%. Given the experimental accuracy and the rounding errors, this is remarkably close agreement.

What are the decay products of a muon at rest? To address this question, we must postulate the following:

1. The energy associated with the rest mass of the muon is $E = mc^2$ (Einstein's mass-energy equivalence equation).

2. The energy for the lifetime ($\sim$2.2 x 10^{-6} seconds) of the muon to exist at rest is $KE_{X4} = -.3mc^2$ (existence equation conjecture).

3. All muons decay to one electron and two neutrinos. Sometimes the decay produces other particles, such as photons.

Based on the above, it can be inferred the rest mass energy ($E = mc^2$) is consumed as follows: $.3mc^2$ is required to satisfy the existence equation conjecture ($E = -.3mc^2$), which gives the muon its at-rest lifetime. The remaining $.7mc^2$ produces the decay particles (electron, two neutrinos, and potentially other particles, such as photons). Although the lifetimes of muons at rest have been accurately measured, the energy of the decay products has not. Therefore, further data is required to validate this inference.

APPENDIX 4

The Philosophy of Time

The title of this appendix is likely to cause most people to yawn. On the surface, it is boring. Therefore, why even include it in the book? What does philosophy have to do with science? The answer to these questions is simple. Your philosophy of time will determine whether you believe time travel is even a scientific possibility. Of the three major philosophical schools on time, only one allows for the possibility of time travel to both the past and future. From this standpoint, it is critical that you know the major philosophies of time and know where you stand on the subject.

Philosophers have been pondering the nature of time for thousands of years. A philosophy of time weaves through almost every ancient culture. For example, the earliest view of the nature of time by a Western philosopher dates back to ancient Egypt and the Egyptian philosopher Ptahhotep (2650–2600 BCE). Indian philosophers and Hindu philosophers also wrote about time dating back to roughly the same period. The ancient Greek philosophers, such as Parmenides,

Heraclitus, and Plato, wrote essays about the nature of time roughly around 500 BCE to 350 BCE.

Many early writers questioned the nature of time, the cause of time, and the unidirectional flow of time, often referred to as the "arrow of time." One of the most interesting aspects when studying the philosophy of time is that some cultures, like the Incas, dating back to about the thirteenth century, considered space and time woven together. Centuries before Einstein published his now-famous special theory of relativity, which scientifically unified space and time (i.e., spacetime), the Incas philosophically unified space and time into a single concept called "pacha."

There is a question about time that has ancient roots and that continues to trouble modern scientists and many religions, namely: Did time have a beginning? Through the ages, philosophers and scientists have struggled with this question, and no widely accepted answer has emerged. Not surprisingly, the "time had no beginning" camp, which originated with the ancient Greeks, held solid ground for over several millennia. The Greeks were formidable philosophers. However, the emerging world religions, including Judaism, Christianity, and Islam, slowly chipped away at the Greek philosophy of an infinite past. They simply taught that a deity made the world, and this suggests a beginning of time. Religious philosophers backed these teachings. Christian philosophers, such as John Philoponus, Muslim philosophers, such as Al-Kindi, and Jewish philosophers, such as Saadia Gaon, argued mathematically that infinities do not exist in reality. If you accept this premise, logically you are backed into a corner and must concede that time had a beginning. In other words, if infinities do not exist in reality and are merely a mathematical construct, then time cannot have an infinite past. This argument was refined and became known as the "argument from the impossibility of completing an actual infinite by successive addition." Simply stated, you cannot complete infinity by adding successive events. Since an infinite past would imply the addition of success events, it ruled out an infinite past. Some notable scientists aligned with this thinking, the most famous today being Stephen Hawking, who argued that time began with the big bang. Dr. Hawking believes that events before the big bang have no observable consequence. It is not clear that this proves time had a beginning. Other physicists, such as Lawrence Krauss, author of *A Universe from Nothing* (2012), and I, author of *Unraveling the Universe's Mysteries* (2012), argue events occurred that preceded and caused the big bang, which implies time preceded the big bang. It does not prove, though, that time has an infinite past or a beginning.

Almost all of us believe we understand time. In fact, when first asked a question about the nature of time, most of us will begin to explain it. However, as we attempt to explain it, the complexity of time's nature emerges. Augustine of Hippo (354 CE–430 CE), known to Christians as St. Augustine, eloquently made this observation: "What then is time? If no one asks me, I know: if I wish to explain it to one that asketh, I know not." The most difficult thing I encountered regarding the nature of time was trying to explain it to my six-year-old grandchild. That is when Einstein's famous quote hit home: "If you can't explain it to a six-year-old, you don't understand it yourself."

Fortunately, though, as the sands of time counted millennia after millennia, three major philosophical schools on the nature of time emerged. Let us examine them and discuss their implications regarding time travel.

Presentism theory of time

The presentism theory of time holds that only the present is real. The past is over. Therefore, it is no longer real. The future has yet to occur. Therefore, the future is not real. Presentists argue that our mind remembers a past and anticipates a future, but neither is real. They are mental constructs.

Arguably, the most famous presentist is Augustine of Hippo (a.k.a. St. Augustine), who compared time to a knife edge. The present represents a knife edge cutting between the past and future. Ironically, this means Augustine of Hippo is not real, since he lived and died in the past. Therefore, should we study Augustine of Hippo, who, by his own philosophy, is not real? Of course, I am only being contentious to make a point.

Presentism has a large following, especially among Buddhists. Fyodor Shcherbatskoy (1866–1942), often referred to as the foremost Western authority on Buddhist philosophy, summed up the Buddhist view of presentism with these few words: "Everything past is unreal, everything future is unreal, everything imagined, absent, mental…is unreal…Ultimately real is only the present moment of physical efficiency." Uncountable millions of Buddhists still ascribe to this view of time and reality.

A cogent philosophical argument can be made for presentism, but presentism is problematic from a scientific viewpoint. When we discussed the special theory of relativity, we learned that the present is a function of the position and speed of the observer. Therefore, what is the present to one observer may be the past to another.

From the standpoint of time travel, presentism renders the question "how to time travel" moot. If we embrace presentism, there is no past or future, and time travel is meaningless. Fortunately, though, other philosophies of time open the door to time travel. Let us examine the next one.

Growing universe theory of time

This theory of time is also referred to as "growing block universe" and "the growing block view." However, regardless of the name, they all hold the same philosophy of time. The past is real, and the present is real. The future is not real. The logic goes something like this: The past is real because it actually happened. We experience it, and we document it. We call it history. The present is real because we experience it. We often share the present with others. The future is not real because it has not occurred.

Why do all the names for this theory of time start with the word "growing"? The concept is that the passage of time continually expands the history of the universe. Actually, this is logical. The history of the world, and the universe, continues to expand with the passage of time. The history section of any library is destined to grow with time.

In this philosophy of time, only time travel to the past makes sense, since for growing-universe philosophers, the past is real. We cannot time travel to the future, since the future has yet to occur. Therefore, it is not real.

As logical as this theory of time may appear, there is scientific evidence that the future is real and can influence the present. We discussed this evidence in the section titled "Twisting the arrow of time" in chapter 1. Now, let us examine the last significant philosophy of time.

Eternalism theory of time

The eternalism theory of time holds that the past, present, and future are real. The philosophy of this theory rests on Einstein's special theory of relativity. Essentially, the special theory of relativity holds that the past, present, and future are functions of the speed and position of an observer.

While Einstein never equated time with the fourth dimension, Minkowski's geometric interpretation of Einstein's special theory of relativity gave birth to four-dimensional space, with time as part of the fourth dimension. In Minkowski's interpretation, often termed "Minkowski space" or "Minkowski spacetime," the fourth dimension includes time and is on equal footing with the normal three-dimensional space we currently encounter. However, Minkowski's fourth dimension borders on the strange. In Minkowski spacetime, the fourth dimension, X_4, is equal to ict, where $i = \sqrt{-1}$, an imaginary number, c is the speed of light in a vacuum, and t is time as measured by clocks. The mathematical expression ict is dimensionally correct, meaning that it is a spatial coordinate, not a temporal coordinate, but is essentially impossible to visualize, since it includes an imaginary number. What is an imaginary number? It is a number that when squared (multiplied by itself) gives a negative number. This is not possible to do with real numbers. If you multiply any real number, even a negative real number like minus one, by itself, you always get a positive number. Therefore, it is impossible to solve for the square root of minus one.

Although we can express it mathematically as $\sqrt{-1}$, it has no solution, and it is termed an imaginary number. Does that mean Minkowski was wrong about the fourth dimension? Actually, it does not. It does say that it is a mathematical construct, and intuitively, for most of us, impossible to visualize. However, the special theory of relativity continues to be taught using Minkowski spacetime, which the bulk of the scientific community considers a valid geometric interpretation. In either its algebraic form, as first presented by Einstein, or its geometric form, as interpreted by Minkowski, the majority of the scientific community considers the special theory of relativity the single most successful theory in science. It has withstood over a century of experimental investigation, and it is widely considered verified.

Because of its scientific underpinnings, the eternalism theory of time is widely accepted in the scientific community. If we adopt the eternalism theory of time, then time travel to the past or future becomes equally valid. In addition, no scientific theory contradicts or prohibits time travel. Said more positively, based on Einstein's theories of relativity, which lay a theoretical foundation for time dilation (i.e., time travel to the future) and closed timelike curves (i.e., time travel to the past), most of the scientific community would support the scientific possibility of time travel.

APPENDIX 5

The Science of Time

From a practical standpoint, the science of time started with Isaac Newton in the seventeenth century, but underwent dramatic changes early in the twentieth century, when a little-known patent examiner published a paper in the *Annalen der Physik* in 1905. The paper contained no references, quoted no authority, and had relatively little in the way of mathematical formulation. The writing style was unconventional for a scientific paper, relying on thought experiments combined with verbal commentary. No one suspected that the world of science was about to be changed forever. The little-known patent examiner was twenty-six-year-old Albert Einstein. The paper was on the special theory of relativity, which quietly led to the scientific unification of space and time, and the scientific realization that mass is equivalent to energy. The ink of this one paper rewrote the science of time. However, we are getting a little bit ahead of ourselves. Let us go back and start with Isaac Newton.

The English physicist Isaac Newton (1642–1727) was the greatest and most famous scientist of his time, and with good reason. He is widely credited with playing a key role in the scientific revolution, hailed as the beginning of modern science. His most important work, the publication of *Philosophiæ Naturalis Principia Mathematica*, Latin for *Mathematical*

Principles of Natural Philosophy, in 1687 set forth his famous three laws of motion (the foundation of Newtonian mechanics), along with his theory of gravity (Newton's law of gravity). Newtonian mechanics and Newton's law of gravity are still taught in high school and college science classes. Newton also contributed to optics and shared the invention (along with Gottfried Leibniz) of calculus, a critical branch of mathematics used in advanced science to this day.

Let us ask the key question: How did Newton scientifically view time? Newton thought of time as an absolute. He believed that time passed uniformly, even in the absence of change. Newton's thoughts about the science of time would go something like this: The world is changing at varying rates, but time passes uniformly. The world stops changing completely, but time passes uniformly. Any event that occurs at a single point in time occurs simultaneously for all observers, regardless of their position or relative motion. Newton's view of time as an absolute became a cornerstone of classical physics and prevailed until the early part of the twentieth century. In our everyday world, this view of time makes complete sense. Newton's science of time only breaks down when observers are at vastly different distances from an event, or when the event or the observers are moving near the speed of light relative to one another. Obviously, this did not occur in the real-world situations of Newton's era. In addition, the speed of light was not a consideration in Newtonian mechanics. Remarkably, Newtonian mechanics is still a highly successful theory for predicting and explaining typical real-world phenomena.

This next part of the story may surprise you. Newton is widely viewed as one of the most influential scientists of all time. His scientific accomplishments and writings make a strong case that his view of time as an absolute was his original work. However, this is probably not entirely true. The concept of time being an absolute actually started with Galileo.

Galileo was a brilliant Italian physicist, mathematician, astronomer, and philosopher. Galileo and Newton never met in person, since

Galileo died the same year Newton was born, 1642. However, Galileo's scientific writings not only played a role in the scientific revolution, but it is likely Galileo played a major role in shaping Newton's thinking. In fact, the coordinate transformation methodology that treats time as an absolute is termed the Galilean transformation. Let us understand how this came about.

Time is an absolute (Galilean transformation)

There appears little doubt that Newton's science of time was significantly influenced by Galileo's 1638 *Discorsi e Dimostrazioni Matematiche* (*Discussions on Uniform Motion*), since Newton's and Galileo's views of time are essentially identical. For example, the transformation of the time coordinate from one frame of reference to another, regardless of the relative motion of either frame, left the time coordinate unchanged. As mentioned above, this type of coordinate transformation is termed the Galilean transformation, and it works as long as the frames of reference move at low velocities. This begs a question. What happens as the frames of reference move at velocities close to the speed of light? To address this question, we need to discuss the Lorenz transformation.

Time is relative (Lorenz transformation)

To transfer the time coordinates between frames moving close to the speed of light, an entirely new transformation methodology needed to be developed. Einstein became painfully aware of this during the development of his famous special theory of relativity. As a result, he utilized the Lorenz transformation. Some authors give Einstein credit as having developed the Lorenz transformation. This is not historically correct. The Lorenz transformation was in existence prior to Einstein's publication of the special theory of relativity in 1905. In fact, numerous physicists, including Voigt (1887), Lorentz (1895), Larmor (1897), and Poincaré (1905), contributed to its formulation. It was Poincaré, in 1905, who brought it into its final modern form and named it the Lorenz transformation. It is fair, though, to say that Einstein rederived the Lorenz transformation in the context of special relativity.

Just what is the Lorenz transformation, and how does it treat the time coordinate between frames moving at a constant rate close to the speed of light? It takes into account that light travels at a finite speed (approximately 186,000 miles/second), and that speed is a constant in any frame of reference moving at a constant velocity, typically referred to as an inertial frame of reference. As a result, according to the Lorenz transformation, different observers moving at different velocities or at rest will not measure time in the same way. Indeed, they may measure different elapsed times, and even a different orderings of events.

Time as a coordinate

At this point, it may appear obvious that time is a coordinate. Both the Galilean and Lorenz transformations view time as a coordinate, and they only differ in how they translate the time coordinate between initial frames of reference. Intuitively, when you think of time as a coordinate, it makes sense. For example, when you set a meeting, you not only set the place (i.e., the spatial coordinates), but the time (i.e., the temporal coordinate). In this context, time is a coordinate (i.e., also known as "coordinate time"). This terminology distinguishes it from "proper time," which is not a coordinate, but rather a process. In Einstein's special theory of relativity, "proper time" (also known as "clock time") is a measure of change, such as the number of rotations of a simple mechanical clock's hands. It is arbitrary. For example, proper time may refer to the time it takes a candle to burn down to a specific point. Before we go further, let us be perfectly clear on the distinction of time as a coordinate

("coordinate time") and proper time ("clock time"). If you specify that you will meet someone at a specific time, you are using "coordinate time." If you say it takes the second hand of your watch one minute to make a compete revolution, you are talking about proper time ("clock time"). You may wonder if coordinate time and proper time are related. It turns out they are. Einstein's special theory of relativity relates coordinate time, proper time, and space to each other via spacetime, which we will discuss next.

Approaching a scientific definition of time

How is coordinate time related to proper time? Einstein's special theory of relativity relates coordinate time and proper time by the following convention. For an observer with a clock in an inertial frame of reference, the coordinate time at the event is equal to proper time at the event when measured by a clock that is stationary relative to the observer and at the same location as the event. This convention assumes synchronization of the clock at the event with the observer's clock. Unfortunately, there have been numerous methods suggested for accurately synchronizing clocks and defining synchronization. For our purposes in defining coordinate time and proper time, it is only important to assume the hands of both clocks move in unison, independent of the method of synchronization.

Most of the scientific community agrees that the most accurate definition of time requires integration with the three normal spatial dimensions (i.e., height, width, and length). Therefore, the scientific community talks in terms of spacetime, especially in the context of relativity, where the event or observer may be moving near the speed of light relative to each another.

Let us consider an example. A clock moving close to the speed of light will appear to run slower to an observer at rest (one frame of reference) relative to the moving clock (another frame of reference). In simple terms, time is not an absolute, but is dependent on the relative motion of the event and observer. It may sound like science fiction that a clock moving at high velocity runs slower, but it is a widely verified science fact. Even the clock on a jet plane flying over an airport will run slightly slower than the clock at rest in the airport terminal. Einstein predicted this time dilation effect in his special theory of relativity, and he derived an equation to calculate the time difference.

Other physical factors affect time. For example, another scientific fact is that a clock in a strong gravitational field will run slower than a clock in a weak gravitational field. Einstein predicted this time dilation effect in his general theory of relativity.

Lastly, the time dilation effects of high velocity and strong gravitational fields are additive. That means a clock moving near the speed of light will move slower if it enters a strong gravitational field.

I termed this section "Approaching a scientific definition of time" for a reason. There is no consensus on the scientific definition of time. However, we can help ourselves conceptualize time by summarizing the salient points discussed above:

- In our everyday existence, time appears to be an absolute. Time is the same for everyone. When an event occurs, we believe it to occur simultaneously regardless of its relative position or velocity to us, or our relative position or velocity to the event. This is our everyday reality. We typically do not worry about accurately synchronizing our watches, since a minute one way or the other does not matter for most real-life applications. The simple fact is that treating time as an absolute works in most real-life applications. However, it is an illusion and breaks down as the relative velocity of any constituent approaches the speed of light, or when the distances from the event become extremely large and different for the observers. On this last point, regarding relative distances, consider the following example. An observer close to an event will record its occurrence a thousand years sooner than an observer a thousand light-years from the event. The reason for this is that it takes the light a thousand years to reach the furthest observer. This means that the stars we view at night may no longer exist. The light has traveled thousands to millions of light-years to reach our eyes. The stars may have died long ago, but we will not know it for thousands to millions of years.

- In the world of relativity, where frames of reference can move near the speed of light or gravitational fields can play a factor, time becomes relative. Here are four examples.

    1. A clock moving near the speed of light will appear to run slower to an observer at rest, relative to the clock.

    2. A clock in a strong gravitational field will appear to run slower to an observer a distance (as little as one meter) farther away from the gravitational field.

    3. A clock will run even slower when moving near the speed of light when it enters a strong gravitational field (i.e., the affects are additive).

    4. An event will appear to occur first to the observer closest to the event. The farther away an observer is from an event, the longer it will take the light to travel to the observer, and for the observer to become aware of the event.

Notice that all our attempts to define time fail. Instead, we describe how time behaves during an interval, a change in time. We are unable to point to an entity and say "that is time." The reason for this is that time is not a single entity, but scientifically an interval. We cannot slice time down to a shadowlike sliver, a dimensionless interval. In fact, scientifically speaking, the smallest interval of time that science can theoretically define, based on the fundamental invariant aspects of the universe, is Planck time.

Planck time is the smallest interval of time that science is able to define. The theoretical formulation of Planck time comes from dimensional analysis, which studies units of measurement, physical constants, and the relationship between units of measurement and physical constants. In simpler terms, one Planck interval is approximately equal to 10^{-44} seconds (i.e., 1 divided by 1 with 44 zeros after it). As far as the science community is concerned, there is a consensus that we would not be able to measure anything smaller than a Planck interval. In fact, the smallest interval science is able to measure as of this writing is trillions of times larger than a Planck interval. It is also widely believed that we would not be able to measure a change smaller than a Planck interval. From this standpoint, we can assert that time is only reducible to an interval, not a dimensionless sliver, and that interval is the Planck interval. Therefore, our scientific definition of time forces us to acknowledge that time is only definable as an interval, the Planck interval.

The time uncertainty interval

Since the smallest unit of time is only definable as the Planck interval, this suggests there is a fundamental limit to our ability to measure an event in absolute terms. This fundamental limit to measure an event in absolute terms is independent of the measurement technology. The error in measuring the start or end of any event will always be at least one Planck interval. This is analogous to the Heisenberg uncertainty principle, which states it is impossible to know the position and momentum of a particle, such as an electron, simultaneously. Based on fundamental theoretical considerations, the scientific community widely agrees that the Planck interval is the smallest measure of time possible. Therefore, any event that occurs cannot be measured to occur less than one Planck interval. This means the amount of uncertainty regarding the start or completion of an event is only knowable to one Planck interval. In our everyday life, our movements consist of a sequence of Planck intervals. We do not perceive this because the intervals are so small that the movement appears continuous, much like watching a movie where the projector is projecting each frame at the rate of approximately sixteen frames per second. Although each frame is actually a still picture of one element of a moving scene, the projection of each frame at the rate of sixteen frames per second gives the appearance of continuous motion. I term this inability to measure an event in absolute terms "the time uncertainty interval."

The relationship between time and existence

Essentially, we can define a mass's movement in time as existence. Let us take a simple example to illustrate this definition. Pretend we are viewing a mass and recording its position, which is at rest. The mass is not moving in any of the spatial dimensions. However, at a later interval (let us pretend our wristwatch records an hour to have passed), we again view the mass and record its position. We observe the mass still exists, and the coordinates are identical to the first set of measurements. In effect, the mass has moved in time along with us. Since the mass and we are in the same frame of reference, we have every reason to believe the rate of the movement of the mass in time is equivalent to our rate of movement in time. If the mass did not move in time—for example, stopped moving in time—it would not be there at the recorded coordinates for the second measurement. We would say it ceased to exist. On this basis, we can assert existence equates to movement in time. In this case, both the mass and we, the observers, moved in time at the same rate.